The Long Story of Earth and People

Azhar ul Haque Sario

Copyright

First Printing, 2025

Azhar.sario@hotmail.co.uk

ORCID: https://orcid.org/0009-0004-8629-830X
Disclaimer: This book is free from AI use. The cover was designed in Microsoft Publisher

Contents

The Paradox of Life's Genesis: A Faint Sun and a Fertile Earth

Subtopic 1: The Young Sun's Dim Glow: Reconstructing early solar luminosity using astrophysical models and stellar evolution data.

A Journey to the Dawn of Time: Unveiling the Secrets of the Faint Young Sun

Imagine our Sun, not as the radiant, life-giving star we know today, but as a dim, nascent celestial body, struggling to ignite its nuclear furnace. This was the reality billions of years ago when our planet was still in its infancy. The young Sun, a mere toddler in astronomical terms, emitted only about 70% of the energy it does today. This presents a perplexing paradox: how could Earth, bathed in such weak sunlight, have maintained a climate warm enough to support liquid water and the emergence of life?

To unravel this enigma, we embark on a journey through time and space, delving into the depths of stellar evolution models and the intricacies of solar irradiance studies. Like cosmic detectives, we piece together clues from the distant past, reconstructing the early solar luminosity and its profound implications for the nascent Earth.

Stellar Evolution: A Tale of Birth, Life, and Death

Stars, like living organisms, undergo a life cycle of birth, growth, maturity, and eventual demise. Our Sun, a G-type main-sequence star, is currently in its middle age, steadily converting hydrogen to helium in its core through nuclear fusion. This process releases vast amounts of energy, radiating outwards as sunlight and sustaining life on Earth.

However, the Sun's energy output has not remained constant throughout its lifetime. As it ages, the Sun gradually becomes brighter and hotter. This is due to the accumulation of helium in its core, which increases its density and temperature, leading to a higher rate of nuclear fusion.

To understand the Sun's early luminosity, we turn to stellar evolution models, sophisticated computer simulations that trace the life cycle of stars from their formation to their ultimate fate. These models incorporate the fundamental laws of physics, including nuclear physics, thermodynamics, and radiative transfer, to predict how stars change over time.

By inputting parameters such as the Sun's initial mass and composition, we can rewind the clock and simulate its evolution backwards, reconstructing its luminosity at different stages of its life. These models reveal that the young Sun was indeed significantly dimmer than it is today, confirming the faint young Sun paradox.

Solar Irradiance Studies: A Window into the Past

While stellar evolution models provide a theoretical framework for understanding the Sun's early luminosity, they are not without limitations. To validate these models and refine our understanding, we turn to solar irradiance studies, which examine the Sun's energy output over time.

One approach involves studying Sun-like stars at different stages of their evolution. By observing these stars, we can gain insights into the Sun's past and future behavior. For instance, the Kepler space telescope has identified numerous Sun-like stars, some younger and some older than our Sun, allowing us to compare their luminosity and activity levels.

Another approach involves analyzing geological records on Earth, such as ice cores and sedimentary rocks, which preserve clues about past climate conditions. These records contain isotopes, chemical elements with varying numbers of neutrons, that can serve as proxies for solar irradiance. For example, the ratio of beryllium-10 to beryllium-9 in ice cores can be used to reconstruct solar activity levels over thousands of years.

By combining these different approaches, we can paint a more complete picture of the Sun's early luminosity and its impact on Earth's climate. The evidence suggests that the young Sun was indeed faint, but Earth's climate remained habitable, thanks to a combination of factors, including a thicker atmosphere and a higher concentration of greenhouse gases.

The Faint Young Sun Paradox: A Puzzle with Profound Implications

The faint young Sun paradox challenges our understanding of the conditions required for life to emerge and thrive. It suggests that Earth's early climate was more resilient than previously thought, capable of maintaining liquid water and supporting life despite receiving less solar energy.

This paradox has profound implications for our search for life beyond Earth. If life could emerge on a planet with a faint young star, then the habitable zone, the region around a star where liquid water can exist on a planet's surface, may be wider than previously thought. This expands the possibilities for finding life on exoplanets, planets orbiting other stars.

The faint young Sun paradox also highlights the interconnectedness of Earth's systems, from the Sun's energy output to the composition of the atmosphere and the dynamics of the oceans. Understanding these complex interactions is crucial for predicting the future of our planet and ensuring its continued habitability.

Subtopic 2: A Greenhouse Cradle: Examining the role of atmospheric gases (CO2, methane, water vapor) in maintaining a habitable climate despite lower solar energy.

Swaddled in Warmth: How Greenhouse Gases Nursed a Young Earth

Picture the early Earth, a volatile world of erupting volcanoes, churning oceans, and a thin, evolving atmosphere. The Sun, a dim ember in the sky, casts a weak glow upon this primordial landscape. Yet, despite the faint solar radiation, Earth's surface remains warm enough to cradle liquid water, the essential ingredient for life. How is this possible?

The answer lies in a delicate balance of atmospheric gases, a gaseous blanket that traps heat and warms the planet, much like a greenhouse. This "greenhouse effect" is a natural phenomenon that has played a crucial role in Earth's climate throughout its history, but it was particularly critical during the early years when the Sun was less luminous.

Let's delve into the key players in this atmospheric drama: carbon dioxide (CO2), methane (CH4), and water vapor

(H_2O). These greenhouse gases, though present in varying concentrations throughout Earth's history, have acted in concert to maintain a habitable climate despite the faint young Sun.

Carbon Dioxide: The Maestro of the Greenhouse Orchestra

Carbon dioxide, a ubiquitous gas in Earth's atmosphere, is often portrayed as a villain in the context of modern climate change. However, in the early Earth, CO_2 played a heroic role, acting as the primary thermostat regulating the planet's temperature.

The early atmosphere was rich in CO_2, possibly hundreds or even thousands of times higher than today's levels. This abundance of CO_2 created a strong greenhouse effect, trapping heat and warming the planet. As the Sun gradually brightened over billions of years, the concentration of CO_2 in the atmosphere decreased, preventing Earth from becoming a scorching inferno.

This delicate balance between solar luminosity and atmospheric CO_2 is a testament to the intricate feedback mechanisms that regulate Earth's climate. As the Sun's energy output increased, the planet's temperature rose, leading to increased weathering of rocks. This weathering process consumed CO_2 from the atmosphere, reducing the greenhouse effect and stabilizing the climate.

Methane: A Potent Greenhouse Gas with a Short Lifespan

Methane, another potent greenhouse gas, also played a significant role in the early Earth's climate. Although less abundant than CO_2, methane is much more effective at trapping heat, molecule for molecule.

The early Earth was likely shrouded in a methane-rich haze, produced by volcanic eruptions and the activity of methanogenic microorganisms. This methane haze would have contributed to the greenhouse effect, further warming the planet.

However, methane has a much shorter lifespan in the atmosphere than CO2. It is readily oxidized, breaking down into CO2 and water. This means that methane's contribution to the greenhouse effect was likely more transient than that of CO2, fluctuating with volcanic activity and microbial activity.

Water Vapor: The Ever-Present Greenhouse Gas

Water vapor, the gaseous form of water, is the most abundant greenhouse gas in Earth's atmosphere today. It is also a highly variable gas, its concentration depending on temperature and other factors.

In the early Earth, water vapor played a crucial role in amplifying the greenhouse effect. As the planet warmed due to CO2 and methane, more water evaporated from the oceans, increasing the concentration of water vapor in the atmosphere. This, in turn, further enhanced the greenhouse effect, creating a positive feedback loop.

Water vapor's role in the early Earth's climate highlights the interconnectedness of Earth's systems. The atmosphere, oceans, and land surface all interact to regulate the planet's temperature and maintain a habitable environment.

Paleoclimate Modeling: Reconstructing the Past to Understand the Present

To unravel the complex interplay of greenhouse gases in the early Earth's climate, scientists rely on paleoclimate modeling, sophisticated computer simulations that reconstruct past climate conditions.

These models incorporate data from various sources, including geological records, ice cores, and fossilized remains, to create a virtual Earth, simulating the interactions between the atmosphere, oceans, and land surface. By adjusting parameters such as atmospheric composition and solar luminosity, scientists can explore different scenarios and test hypotheses about the early Earth's climate.

Paleoclimate models have provided valuable insights into the role of greenhouse gases in maintaining a habitable climate despite the faint young Sun. They have shown that a combination of CO2, methane, and water vapor, along with other factors such as cloud cover and volcanic activity, could have kept the early Earth warm enough for liquid water to exist.

A Delicate Balance: Lessons for the Future

The story of Earth's early climate is a tale of delicate balance, a testament to the intricate interplay of atmospheric gases and solar radiation. It highlights the importance of understanding the greenhouse effect, not just as a driver of modern climate change, but as a fundamental process that has shaped Earth's history and enabled life to flourish.

As we face the challenges of a changing climate today, the lessons from the early Earth are more relevant than ever. By studying the past, we can gain insights into the complex feedback mechanisms that regulate our planet's climate and develop strategies to mitigate the impacts of human activities.

The early Earth, swaddled in a greenhouse cradle, reminds us that our planet's climate is a precious and fragile gift, one that we must strive to protect for future generations.

Subtopic 3: The Crucible of the Oceans: Exploring the role of hydrothermal vents, volcanic activity, and early oceans in providing the chemical building blocks and energy for life's emergence.

In the Depths of the Primordial Sea: Where Life First Stirred

Imagine the early Earth's oceans, a vast, churning expanse of water, teeming with dissolved minerals and volcanic gases. Beneath the surface, a hidden world of hydrothermal vents spews forth hot, mineral-rich fluids, creating chemical gradients and energy flows that would spark the genesis of life.

This is the crucible of the oceans, a realm where the boundaries between the geological and the biological blur. Here, in the depths of the primordial sea, the stage was set for the emergence of life, a story written in the language of chemistry and thermodynamics.

Hydrothermal Vents: Oases of Life in the Deep

Hydrothermal vents, discovered in the late 1970s, are fissures in the ocean floor where geothermally heated water erupts. These vents, often found near volcanically active zones, create unique chemical environments that support thriving ecosystems, independent of sunlight.

The vents release a cocktail of chemicals, including hydrogen sulfide, methane, and iron, which serve as universities are preparing people for these incredible careers.

This isn't science fiction anymore. The $400 billion space economy, powered by relentless private energy and investment, is generating real value, diverse jobs, and mind-bending innovation today. The grand challenge, and the immense opportunity, is to intelligently weave this powerful new thread into our national economies, ensuring that humanity's next giant leap outward translates directly into tangible progress for everyone back home. The stars are open for business, and we're all invited to participate in building that future.

The Cosmic Crossroad: Navigating the New Rush Beyond Earth

Look up. The quiet expanse isn't so quiet anymore. Humanity's reaching for the stars again, but this time, it's not just superpowers planting flags. It's a buzzing, jostling crowd: dreamers building rockets in garages, billionaires charting courses to the Moon, companies promising orbital vacations and asteroid fortunes. It's exhilarating, a true frontier opening up. But like any frontier, it's wild, untested, and we're scrambling to write the rules of the road while the traffic is already speeding by. Our old maps – the regulations drawn up when space was mostly empty – are struggling to guide this new, complex cosmic dance.

Think about the launchpads right here on Earth, especially in the US. Imagine the engineers, fueled by coffee and ambition, finally ready to send their creation skyward. But first, they hit the regulatory queue. The FAA, the designated traffic cops for space launches, are facing a deluge. Picture stacks of complex applications from SpaceX's star-hopping ambitions, Blue Origin's tourist trips, Rocket Lab's workhorse vehicles, and countless smaller players dreaming big. It's not just paperwork; it's a fundamental tension wire humming between "Let's innovate!" and "Let's be absolutely sure no one gets hurt." As legal eagles writing in space journals point out, these delays aren't just inconvenient; they can stall dreams, scare off investors, and make it harder for the little guys to even get into the game. The conversation isn't just about more funding for the FAA; it's about finding smarter, more flexible ways to ensure safety without strangling the very innovation that's pushing us outward. Maybe it's less about prescribing how to be safe, and more about defining what safety looks like, letting these bright minds figure out the best way to get there.

Now, zoom out, way out, to the global stage. We have this foundational document, the Outer Space Treaty of 1967. Born from Cold War anxieties, it was remarkably wise for its time, declaring space open to all, forbidding nations from claiming the Moon like conquistadors, and making countries responsible for everything launched from their soil. But 1967 couldn't imagine today. As legal minds exploring Air and Space Law regularly debate: What happens when a company, not a country, starts digging up lunar regolith? Is that 'claiming' it? And who's liable when a piece of debris from a long-dead Soviet satellite hits a brand-new commercial one launched by a multinational consortium? The Treaty offers grand principles, like a constitution, but lacks the specific bylaws for cosmic resource management, orbital traffic jams, and cleaning up our celestial backyard – things critical to today's space reality. This fuzziness creates risks, the kind that could spark real conflict as asteroid mining shifts from sci-fi fantasy to boardroom strategy. Efforts like the Artemis Accords are trying to fill these gaps, particularly on using space resources, but it's like a group agreeing on house rules while others aren't part of the conversation – we need everyone on the same page.

Regardless of the specific mechanism, the emergence of life was a remarkable event, a testament to the creative power of chemistry and the resilience of life in the face of adversity.

Extremophiles: Clues to Life's Origins

Extremophiles, organisms that thrive in extreme environments, offer clues to the conditions under which life may have first emerged. These hardy microbes inhabit environments ranging from boiling hot springs to acidic mine drainage, demonstrating the remarkable adaptability of life.

Some extremophiles, such as thermophiles and hyperthermophiles, thrive in high temperatures, similar to those found in hydrothermal vents. These organisms provide insights into the potential for life to exist in extreme environments, both on Earth and beyond.

Other extremophiles, such as acidophiles and alkaliphiles, tolerate extreme pH levels, suggesting that life could have emerged in acidic or alkaline environments, such as those found near volcanic vents.

By studying extremophiles, we can gain a better understanding of the limits of life and the potential for life to exist in diverse environments throughout the universe.

A Continuing Journey of Discovery

The exploration of the early oceans and the origins of life is a continuing journey of discovery. New findings and technologies are constantly expanding our understanding of this fascinating chapter in Earth's history.

From the depths of hydrothermal vents to the vast expanse of the open ocean, the crucible of life continues to reveal its secrets, offering glimpses into the remarkable processes that gave rise to life on our planet.

Subtopic 4: Life's Tenacity: Lessons from Extremophiles: Investigating extremophile organisms thriving in extreme environments (e.g., deep-sea vents, acidic hot springs) as analogs for early life on Earth.

Life on the Edge: Extremophiles and the Limits of Possibility

Venture into the most inhospitable corners of our planet – the scalding heat of volcanic springs, the crushing depths of the ocean floor, the arid desolation of salt flats – and you'll find life, not just surviving, but thriving. These are the extremophiles, organisms that laugh in the face of conditions that would obliterate most other life forms. Their existence offers a profound testament to life's tenacity and provides invaluable clues to the origins and potential for life beyond Earth.

Extremophiles are not merely curiosities; they are windows into the past, offering glimpses of what early life on Earth might have been like. They challenge our anthropocentric view of what constitutes a "habitable" environment and expand our understanding of life's potential to adapt and persist in the face of adversity.

Masters of Adaptation: Thriving in the Extremes

Extremophiles have evolved a dazzling array of strategies to cope with their challenging environments. Thermophiles, for instance, thrive in scorching temperatures, their proteins and enzymes specially adapted to function in the heat. Psychrophiles, on the other hand, flourish in frigid conditions, their cell membranes remaining fluid even in sub-zero temperatures.

Halophiles are salt-loving organisms, inhabiting hypersaline environments like the Dead Sea. They have evolved mechanisms to maintain osmotic balance, preventing their cells from shriveling up in the high salt concentrations. Acidophiles thrive in acidic environments, such as volcanic hot springs and acid mine drainage, while alkaliphiles prefer alkaline conditions, like soda lakes.

These are just a few examples of the incredible diversity of extremophiles. Each group has evolved unique adaptations to their specific environment, showcasing the remarkable plasticity of life.

Deep-Sea Vents: Echoes of the Early Earth

Deep-sea hydrothermal vents, discovered only in the late 20th century, are perhaps the most compelling analogs for the early Earth. These vents, spewing forth hot, mineral-rich fluids from the Earth's interior, support thriving ecosystems independent of sunlight.

The organisms that inhabit these vents, such as tube worms and giant clams, rely on chemosynthetic bacteria for sustenance. These bacteria harness the chemical energy from the vent fluids, converting inorganic compounds into organic matter, forming the base of the food chain.

The conditions at deep-sea vents – high temperatures, high pressure, and a lack of sunlight – are thought to resemble those of the early Earth. The discovery of life thriving in these environments has fueled speculation that life may have originated in similar settings, perhaps around hydrothermal vents.

Acidic Hot Springs: A Crucible for Life?

Acidic hot springs, another extreme environment, also offer insights into the potential for life to emerge in harsh conditions. These springs, often associated with volcanic activity, are characterized by low pH, high temperatures, and a rich mix of dissolved minerals.

Despite these challenges, acidic hot springs teem with life, from colorful microbial mats to specialized insects and crustaceans. The organisms that inhabit these environments have evolved adaptations to tolerate the acidity and heat, offering clues to the early evolution of life.

Some scientists believe that acidic hot springs may have been the cradle of life, providing a unique environment for the first self-replicating molecules to emerge. The combination of heat, acidity, and dissolved minerals could have created a chemical cauldron, driving the formation of organic molecules and the emergence of life.

Extremophiles and Astrobiology: Expanding the Search for Life

The study of extremophiles has profound implications for astrobiology, the search for life beyond Earth. By understanding the limits of life on Earth, we can better assess the potential for life to exist on other planets and moons.

Extremophiles have expanded our definition of what constitutes a "habitable" environment. They have shown that life can thrive in conditions that were once thought to be inhospitable, suggesting that life may exist in a wider range of environments than previously thought.

For example, the discovery of extremophiles thriving in the dry valleys of Antarctica, one of the most Mars-like environments on Earth, has fueled hopes of finding life on the Red Planet. Similarly, the existence of extremophiles in subglacial lakes, such as Lake Vostok in Antarctica, has raised the possibility of life existing beneath the icy surface of Europa, one of Jupiter's moons.

Lessons from the Edge: The Resilience of Life

Extremophiles offer a powerful testament to the resilience of life. They demonstrate that life can adapt and thrive in the most challenging environments, pushing the boundaries of what we thought possible.

By studying these remarkable organisms, we gain a deeper appreciation for the diversity and tenacity of life. We learn that life is not confined to a narrow range of conditions, but can flourish in a multitude of environments, from the depths of the ocean to the heights of the atmosphere.

Extremophiles remind us that life is a force to be reckoned with, a force that can overcome seemingly insurmountable obstacles. They inspire us to explore the limits of life, both on Earth and beyond, and to continue the search for life in all its forms.

A Final Word

The faint young Sun paradox, a puzzle that has challenged our understanding of Earth's early history, has led us on a journey through time and space, from the depths of the oceans to the heights of the atmosphere. We have explored the role of greenhouse gases, hydrothermal vents, and volcanic activity in shaping the early Earth's environment and creating the conditions for life to emerge.

We have also delved into the fascinating world of extremophiles, organisms that thrive in the most extreme environments on Earth. These remarkable creatures offer clues to the origins of life and expand our understanding of the potential for life to exist beyond Earth.

The story of the faint young Sun paradox is a testament to the interconnectedness of Earth's systems and the remarkable resilience of life. It is a story that continues to unfold, as new discoveries and technologies reveal more about our planet's history and the origins of life.

Zircon Whispers: Decoding Earth's First Billion Years

Subtopic 1: Time Capsules in Stone: Exploring the unique properties of zircon crystals that make them ideal for preserving information about Earth's early history.

Imagine holding a tiny grain of sand, seemingly insignificant, yet within its crystalline structure lies a story that stretches back billions of years to the dawn of our planet. This is the magic of zircon, a mineral that has become the geologist's ultimate time machine.

Zircons are crystals composed of zirconium silicate ($ZrSiO4$), and their resilience is legendary. They are born from the fiery depths of volcanoes, forged under immense pressure and heat. This dramatic origin gives them incredible durability, allowing them to survive the relentless forces of erosion, metamorphism, and even the melting of their parent rocks. Like microscopic fortresses, they guard their secrets across the vast expanse of geologic time.

But zircon's timekeeping abilities go beyond mere survival. Their crystal structure has a remarkable affinity for uranium, readily incorporating it during their formation. Uranium, being radioactive, decays into lead at a predictable rate. This is the key to unlocking zircon's chronological secrets. By measuring the ratio of uranium to lead within a zircon crystal, scientists can determine its age with astonishing precision. This method, known as U-Pb dating, has revolutionized our understanding of Earth's history.

The beauty of zircon as a time capsule lies not only in its ability to record its own birthdate but also in its capacity to capture snapshots of its surroundings. Like a microscopic fly trapped in amber, zircons can encapsulate other minerals, tiny droplets of magma, and even traces of ancient seawater during their growth. These inclusions provide invaluable clues about the conditions prevailing on Earth when the zircon crystallized.

For instance, the oxygen isotopes trapped within a zircon can reveal the temperature of the water from which it formed, offering insights into the early oceans. Similarly, the presence of certain rare earth elements can provide clues about the composition of the magma from which the zircon crystallized, shedding light on the early Earth's crustal evolution.

Zircons from the Jack Hills region of Western Australia, some dating back an astonishing 4.4 billion years, have provided tantalizing evidence for the existence of liquid water on Earth's surface just a few hundred million years after its formation. This discovery has profound implications for our understanding of the early Earth's environment and the potential for the emergence of life.

In essence, zircons are not just crystals; they are miniature time capsules, each one a storybook waiting to be deciphered. Their resilience, their affinity for uranium, and their capacity to capture fragments of their environment make them the ultimate tool for piecing together the enigmatic history of our planet's first billion years.

Subtopic 2: Unlocking Deep Time: Explaining the principles of radiometric dating (U-Pb method) and how it's applied to zircons to determine their ages and the conditions of their formation.

Imagine a clock that ticks not with seconds, but with millennia. This is the essence of radiometric dating, a technique that harnesses the predictable decay of radioactive elements to measure the vast stretches of geologic time. And within the realm of radiometric dating, the U-Pb method applied to zircons reigns supreme, offering unparalleled accuracy and precision in unraveling Earth's ancient past.

At the heart of this method lies the radioactive decay of uranium. Uranium exists as two primary isotopes: uranium-238 (238U) and uranium-235 (235U). Both isotopes are unstable and undergo a series of decay steps, ultimately transforming into stable isotopes of lead – lead-206 (206Pb) and lead-207 (207Pb), respectively. The magic lies in the fact that this decay occurs at a constant, known rate, unaffected by external factors like temperature or pressure.

This is where zircons come into play. As they crystallize, they readily incorporate uranium into their structure, effectively setting the clock ticking. Over time, the uranium within the zircon decays into lead. By measuring the ratio of parent uranium to daughter lead isotopes in a zircon crystal, scientists can calculate the time elapsed since its formation – its age.

The U-Pb method is particularly powerful because it utilizes two independent decay chains (238U to 206Pb and 235U to 207Pb). This provides a built-in cross-check, enhancing the accuracy and reliability of the age determination. Moreover, advancements in analytical techniques, such as mass spectrometry, have allowed scientists to measure these isotopic ratios with incredible precision, pushing the boundaries of geochronology.

But the information gleaned from zircons goes beyond just their age. The U-Pb system can also reveal the conditions under which the zircon crystallized. For instance, the presence of discordant ages, where the ages derived from the two uranium-lead decay chains don't agree, can indicate that the zircon has experienced a thermal event, such as metamorphism, since its formation. By analyzing these discrepancies, scientists can reconstruct the thermal history of the rock and gain insights into the tectonic forces that have shaped it.

Furthermore, the U-Pb system can be combined with other isotopic systems, such as the lutetium-hafnium (Lu-Hf) system, to provide even richer information about the zircon's origin. The Lu-Hf system, which also involves radioactive decay, can help trace the source of the magma from which the zircon crystallized, providing clues about the evolution of Earth's crust and mantle.

In essence, the U-Pb dating of zircons is like having a conversation with the very rocks themselves, allowing us to decipher their stories and piece together the grand narrative of Earth's history. It's a testament to the power of human ingenuity to unlock the secrets hidden within these tiny time capsules and illuminate the deep time that stretches back billions of years.

Subtopic 3: The Hadean Eon Unveiled: Analyzing the insights gained from Hadean zircons (older than 4 billion years) about the early Earth's crust, oceans, and potential for life.

Journey back with me to the Hadean Eon, a time over 4 billion years ago, when Earth was a fiery, volatile world, still reeling from its violent birth. This period, shrouded in mystery due to the scarcity of surviving rocks, is like a faded photograph of our planet's infancy. Yet, within the resilient crystals of Hadean zircons, we find astonishingly clear glimpses of this enigmatic era.

These ancient zircons, primarily discovered in the Jack Hills of Western Australia, are like time travelers from a lost world. They whisper tales of a surprisingly dynamic early Earth, challenging the traditional view of a hellish, molten wasteland.

One of the most profound revelations from Hadean zircons is the evidence for the existence of liquid water on Earth's surface remarkably early in its history. Oxygen isotope analysis of these zircons suggests they crystallized in a relatively cool environment, implying the presence of oceans, perhaps even as early as 4.4 billion years ago. This has profound implications for the emergence of life, as water is considered a crucial ingredient for its development.

Furthermore, the chemical composition of Hadean zircons provides clues about the nature of the early Earth's crust. Many of these zircons show characteristics indicative of formation in granitic rocks, suggesting that the processes of plate tectonics, which generate continental crust, may have been active much earlier than previously thought. This challenges the traditional view of a primordial Earth covered in a global magma ocean and suggests a more differentiated planet with continents and oceans already taking shape.

The inclusions trapped within Hadean zircons also offer tantalizing hints about the early Earth's environment. Minerals like muscovite and quartz suggest the presence of a hydrated crust, further supporting the idea of early oceans. The discovery of graphite inclusions in some Hadean zircons has even sparked debate about the possibility of early life, although this remains a controversial topic.

However, interpreting the whispers of Hadean zircons is not without its challenges. Their scarcity and the potential for alteration over billions of years require careful analysis and interpretation. Scientists are constantly developing new techniques and approaches to extract the maximum information from these precious time capsules.

Despite these challenges, the study of Hadean zircons has revolutionized our understanding of the early Earth. They have painted a picture of a dynamic and evolving planet, with oceans, continents, and perhaps even the potential for life emerging much earlier than previously imagined. As we continue to explore these ancient crystals, we are gradually unveiling the secrets of Earth's first billion years, a period that holds the key to understanding our planet's origins and the emergence of life itself.

Subtopic 4: Continental Genesis: Tracing the growth and evolution of continents through time using zircon data and its implications for the distribution of resources and the development of life.

Continents, the vast landmasses that cradle our civilizations, are not static entities. They are dynamic, ever-changing features of our planet, shaped by the relentless forces of plate tectonics over billions of years. And within the tiny crystals of zircons, we find a remarkable record of this continental evolution, a story that intertwines with the distribution of resources and the very development of life.

Zircons, with their incredible resilience and ability to preserve information about their formation, act as "fingerprints" of the continents. By analyzing the ages and compositions of zircons from different regions, scientists can trace the growth and evolution of continents through time, piecing together the puzzle of their origins.

One of the key insights from zircon studies is that continents did not form all at once. Instead, they have grown episodically through a process called crustal differentiation. This process involves the partial melting of the Earth's mantle, generating magmas that rise and solidify to form new continental crust. Zircons, crystallizing from these magmas, capture the chemical signature of their source, allowing scientists to track the evolution of the continental crust through time.

The earliest continents, forming in the Hadean and Archean Eons, were likely small and fragmented, resembling volcanic island arcs. Over time, these proto-continents collided and amalgamated, gradually assembling into larger landmasses. Zircons from these ancient terrains record this episodic growth, revealing periods of intense magmatism and crustal formation.

The growth of continents has profound implications for the distribution of resources. Many valuable mineral deposits, such as gold, copper, and diamonds, are associated with specific geological processes that occur during continental formation and evolution. By understanding the history of continents, as revealed by zircon data, we can better predict the location of these resources, guiding exploration efforts and ensuring their sustainable use.

Moreover, the evolution of continents has played a crucial role in the development of life. The weathering of continental rocks provides essential nutrients to the oceans, fueling the growth of marine organisms. The emergence of large continents also created diverse terrestrial environments, driving the evolution of land plants and animals. Zircons, by tracing the history of continents, provide a backdrop for understanding the co-evolution of Earth's lithosphere and biosphere.

However, the zircon record of continental evolution is not without its biases. The preservation of zircons is influenced by various factors, including erosion, burial, and tectonic processes. This can lead to an incomplete picture of continental history, particularly for the earliest periods. Scientists are constantly developing new techniques and approaches to address these biases and refine our understanding of continental genesis.

In essence, zircons are not just timekeepers; they are storytellers of continental evolution. By deciphering their chemical and isotopic codes, we can trace the growth of continents, understand the distribution of resources, and appreciate the profound influence of continents on the development of life. As we continue to explore these tiny crystals, we are unraveling the grand narrative of Earth's dynamic history, a story written in stone and whispered by zircons.

Cosmic Crucible: The Late Heavy Bombardment and Life's Resilience

Subtopic 1: Scars of Creation: Examining the evidence for the LHB from lunar craters, impact basins on Earth, and the ages of meteorites.

The Late Heavy Bombardment (LHB) is a hypothesized period of intense asteroid and cometary impacts in the early Solar System, roughly 4.1 to 3.8 billion years ago. While it's a widely accepted theory, definitive evidence remains somewhat elusive. However, we can piece together a compelling picture of this chaotic period by examining the "scars" left behind on celestial bodies and within the very rocks that bombard our planet.

Lunar Craters: A Chronicle of Impacts

The Moon, our closest celestial neighbor, serves as a remarkable archive of the Solar System's early history. Unlike Earth, the Moon lacks the erosive forces of wind, water, and plate tectonics, preserving a detailed record of impact events.

Crater Counting: The lunar surface is pockmarked with craters of all sizes, from microscopic to vast basins hundreds of kilometers across. By analyzing the size distribution and density of these craters, scientists can infer the relative ages of different lunar regions. Older surfaces have accumulated more craters over time.

Impact Basins: The Moon also bears the scars of colossal impacts that formed massive basins, such as the Imbrium Basin, which spans over 1,100 kilometers. These basins are often surrounded by rings of mountains and filled with lava flows, providing further clues about the timing and intensity of the LHB.

Apollo Samples: The Apollo missions brought back hundreds of kilograms of lunar rocks, including samples of impact melt – rocks melted by the heat of impacts. Radiometric dating of these samples has revealed a clustering of ages around 3.9 billion years ago, suggesting a surge in impact activity during that period.

Earthly Scars: Hidden in Plain Sight

While Earth's active geology has erased much of its early impact record, evidence of the LHB can still be found:

Ancient Craters: A few ancient impact craters have survived on Earth, such as the Vredefort Crater in South Africa (300 km diameter) and the Sudbury Basin in Canada (250 km diameter). Although eroded and deformed, these structures provide evidence of the immense forces unleashed during the LHB.

Impact Spherules: Microscopic glass beads called impact spherules are found in ancient rock layers. These spherules formed when molten rock ejected from impacts cooled and solidified in the atmosphere. Their presence in layers dating back to the LHB era supports the idea of a period of heightened impact activity.

Meteorites: Messengers from the Past

Meteorites, fragments of asteroids and comets that fall to Earth, offer another window into the LHB.

Chronometers of the Solar System: Many meteorites contain minerals that can be dated radiometrically. By analyzing these minerals, scientists have found that a significant number of meteorites show evidence of shock metamorphism – changes in their structure caused by high-pressure impacts – around 3.9 billion years ago. This further strengthens the case for the LHB.

Types of Meteorites: The types of meteorites that fell to Earth during the LHB may also provide clues about the source of the impacting objects. For instance, an abundance of iron meteorites could suggest that collisions in the asteroid belt were responsible for the LHB.

Challenges and Uncertainties

Despite the evidence, the LHB remains a topic of ongoing research and debate. Some scientists argue that the apparent clustering of impact ages could be a sampling bias or the result of a single, massive impact event. Others question whether the LHB was a truly distinct period or simply a tail end of the early Solar System's accretion process.

Looking Ahead

Future missions to the Moon and Mars, along with continued analysis of lunar samples and meteorites, are expected to shed more light on the LHB. These studies will help us better understand the early history of our Solar System and the conditions under which life emerged on Earth.

Subtopic 2: Cataclysm and Renewal: Analyzing the potential environmental consequences of the LHB on early Earth, including atmospheric changes, ocean vaporization, and potential mass extinctions.

Imagine a young Earth, its surface still molten in places, its atmosphere a hazy mix of volcanic gases. Now picture a relentless barrage of asteroids and comets, each impact unleashing unimaginable energy, scarring the planet and transforming its environment in profound ways. This is the picture painted by the Late Heavy Bombardment (LHB), a period of intense cosmic upheaval that may have had dramatic consequences for early Earth.

Atmospheric Upheaval:

Vaporization and Shockwaves: Each impact would have vaporized vast quantities of rock and water, sending shockwaves rippling through the atmosphere. These shockwaves could have heated the atmosphere to scorching temperatures, potentially sterilizing the surface.

Atmospheric Erosion: The energy from impacts could have stripped away portions of Earth's early atmosphere, potentially altering its composition and affecting its ability to retain heat.

Greenhouse Effect: The release of gases like carbon dioxide and water vapor from impacts could have intensified the greenhouse effect, leading to periods of extreme warming.

Oceanic Disruption:

Vaporization: Large impacts could have vaporized entire oceans, creating a steamy atmosphere that would have taken considerable time to cool and condense back into liquid water.

Tsunamis: Mega-tsunamis, dwarfing anything seen in modern times, would have ravaged coastlines, eroding landmasses and reshaping the landscape.

Ocean Chemistry: The influx of material from impacting asteroids and comets could have altered the chemistry of the oceans, potentially affecting the development of early life.

Mass Extinctions:

Sterilization Events: The intense heat, pressure, and radiation from impacts could have created uninhabitable conditions for early life, leading to localized or even global extinction events.

Ecological Disruption: Even if life survived the initial impacts, the long-term environmental consequences of the LHB, such as atmospheric changes and ocean acidification, could have disrupted ecosystems and driven further extinctions.

A Crucible for Life:

Paradoxically, the LHB may have also played a role in the emergence and diversification of life.

Hydrothermal Vents: Impacts could have created hydrothermal vent systems, providing energy and nutrients for early life forms.

Chemical Evolution: The delivery of organic molecules and minerals from impacts could have contributed to the chemical building blocks necessary for life.

Evolutionary Pressure: The harsh conditions of the LHB may have favored the survival of extremophiles – organisms that thrive in extreme environments – potentially setting the stage for the evolution of more complex life forms.

Case Studies:

The Chicxulub Impact: The impact that wiped out the dinosaurs 66 million years ago provides a glimpse into the devastating effects of a large impact event. Although much smaller than the impacts during the LHB, the Chicxulub impact caused widespread wildfires, tsunamis, and a global impact winter, leading to the extinction of 76% of plant and animal species.

The Sudbury Basin: This impact crater in Canada is one of the oldest and largest on Earth, dating back to the LHB era. Studies of the Sudbury Basin have revealed evidence of massive melting and vaporization of rock, as well as the release of large amounts of sulfur dioxide into the atmosphere, which could have caused acid rain and global cooling.

Looking Ahead:

Understanding the environmental consequences of the LHB is crucial for understanding the early evolution of Earth and the emergence of life. Future research, including computer simulations and the study of ancient rocks, will help us better understand the complex interplay between impacts, climate, and life during this tumultuous period.

Subtopic 3: Seeds of Life from the Stars: Exploring the possibility that asteroid and comet impacts delivered water, organic molecules, and even the building blocks of life to early Earth.

The origin of life on Earth is one of the most fundamental and captivating mysteries in science. While we've made significant strides in understanding the processes that could have led to the first self-replicating molecules, the question of where the necessary ingredients came from remains open. One intriguing possibility is that these vital components were delivered to Earth by celestial messengers – asteroids and comets – during the early bombardment phase of our planet's history.

Water Delivery:

Earth's Dry Start: Current models suggest that Earth may have formed in a relatively dry region of the solar system, with limited access to water.

Hydrated Asteroids and Comets: Asteroids and comets, especially those originating from the outer regions of the solar system, often contain significant amounts of water ice.

Impact Delivery: During the LHB, countless asteroids and comets bombarded Earth. These impacts could have deposited vast quantities of water, contributing to the formation of our oceans.

Isotopic Evidence: Analysis of the isotopic composition of water in Earth's oceans and in samples from comets and asteroids can provide clues about their potential relationship. While there are some discrepancies, certain types of asteroids, like carbonaceous chondrites, show isotopic similarities to Earth's water, supporting the idea of an extraterrestrial origin.

Organic Molecules:

Building Blocks of Life: Organic molecules, such as amino acids, nucleobases, and sugars, are the essential building blocks of life as we know it.

Extraterrestrial Organics: Meteorites, particularly carbonaceous chondrites, have been found to contain a wide array of organic molecules. These molecules are thought to have formed in the interstellar medium or within the protoplanetary disk from which our solar system formed.

Impact Delivery and Survival: During the LHB, these organic-rich meteorites could have delivered their precious cargo to Earth. While some molecules may have been destroyed during the fiery entry and impact, others could have survived, especially if they were embedded within the meteorite's interior.

Case Study: The Murchison Meteorite: This meteorite, which fell in Australia in 1969, is a treasure trove of organic molecules, including amino acids, purines, pyrimidines, and sugars. The diversity and complexity of these molecules highlight the potential for extraterrestrial delivery of life's building blocks.

Beyond Organics:

Prebiotic Chemistry: Some researchers propose that even more complex molecules, such as peptides and lipids, could have formed in space and been delivered to Earth. These molecules could have jump-started prebiotic chemistry, the chemical processes that preceded the emergence of life.

Chirality: One of the intriguing aspects of life on Earth is the prevalence of left-handed amino acids and right-handed sugars. Some meteorites have been found to contain an excess of one chiral form over the other, suggesting that the bias towards specific chirality may have originated in space.

Challenges and Future Directions:

Survival of Organics: One of the main challenges is understanding how delicate organic molecules could have survived the harsh conditions of atmospheric entry and impact.

Delivery Efficiency: It's also important to determine how efficiently these molecules could have been delivered to Earth and incorporated into environments suitable for life.

Further Research: Future missions to asteroids and comets, along with continued analysis of meteorites and laboratory experiments simulating early Earth conditions, will help us better understand the role of extraterrestrial delivery in the origin of life.

The idea that life's ingredients were seeded from the stars is both humbling and inspiring. It suggests that life may be more widespread in the universe than we once thought, and that the seeds of life may be scattered throughout the cosmos, waiting for the right conditions to take root.

Subtopic 4: Life's Endurance: Investigating how life not only survived the LHB but may have thrived in its aftermath, leading to increased diversification and evolutionary innovation.

The Late Heavy Bombardment (LHB) paints a picture of early Earth as a chaotic, volatile world, seemingly inhospitable to life. Yet, life not only persisted through this tumultuous period but may have even thrived, emerging from the crucible of impacts more diverse and resilient than before. This subtopic delves into the remarkable story of life's endurance during the LHB, exploring the strategies it employed to survive and the ways in which the bombardment may have inadvertently spurred evolutionary innovation.

Survival Strategies:

Extremophiles: The early Earth was already home to extremophiles – organisms capable of surviving in extreme environments, such as those with high temperatures, acidity, or radiation. These hardy microbes were well-suited to withstand the harsh conditions of the LHB.

Subsurface Refugia: Deep beneath the Earth's surface, life could have found refuge from the impacts. Subterranean environments, such as hydrothermal vents and deep-sea trenches, would have been shielded from the most destructive effects of the bombardment.

Endospores: Some bacteria can form endospores, dormant, highly resistant structures that can survive for extended periods in harsh conditions. Endospores could have allowed microbes to "weather the storm" of the LHB, re-emerging when conditions became more favorable.

Horizontal Gene Transfer: The exchange of genetic material between different microbial species, known as horizontal gene transfer, may have played a crucial role in adaptation. This process allows microbes to quickly acquire new traits, potentially conferring resistance to the stresses imposed by the LHB.

Evolutionary Innovation:

Diversification: The LHB may have acted as a selective pressure, favoring organisms with traits that enhanced their survival. This could have led to an increase in the diversity of life, as different lineages adapted to different niches and challenges.

New Metabolic Pathways: The environmental changes caused by the LHB, such as the release of nutrients and the creation of new habitats, may have spurred the evolution of novel metabolic pathways. For instance, the influx of sulfur from impacts could have led to the development of sulfur-based metabolism.

The Rise of Eukaryotes: Some scientists speculate that the LHB may have played a role in the emergence of eukaryotes, the complex cells that make up all plants and animals. The stress of the bombardment could have favored the evolution of more sophisticated cellular structures and mechanisms.

Evidence and Case Studies:

Early Microbial Fossils: Fossil evidence suggests that life was present on Earth as early as 3.5 billion years ago, during the LHB. These early microbes exhibit a surprising degree of diversity, indicating that life had already begun to diversify and adapt.

Genetic Studies: Analysis of the genomes of modern organisms can provide insights into the evolutionary history of life. Studies have revealed evidence of ancient gene transfer events and adaptations that may be linked to the LHB.

The Chicxulub Impact and Mammalian Evolution: The Chicxulub impact, while not part of the LHB, provides an example of how a catastrophic event can lead to evolutionary opportunities. The extinction of the dinosaurs paved the way for the diversification and rise of mammals.

The LHB: A Catalyst for Life?

While often portrayed as a destructive force, the LHB may have ultimately been a catalyst for life's evolution. By challenging early life with a series of environmental upheavals, the bombardment may have inadvertently driven the development of greater resilience, diversity, and complexity. This perspective challenges the traditional view of impacts as solely destructive events, highlighting the potential for even the most catastrophic events to shape the trajectory of life.

Conclusion:

The story of life's endurance through the LHB is a testament to its remarkable adaptability and resilience. Not only did life survive the bombardment, but it emerged stronger and more diverse, setting the stage for the evolution of all the life we see today. This chapter in Earth's history reminds us that even in the face of seemingly insurmountable challenges, life finds a way to persist and even thrive.

Beyond the City Walls: Redefining "Civilization" in the Anthropocene

Subtopic 1: The Limits of Traditional Definitions: Critiquing the conventional markers of civilization (urbanization, agriculture, writing) and their limitations in capturing the diversity of human societies.

Beyond Brick and Mortar: Rethinking "Civilization" in the 21st Century

The term "civilization" has long been a loaded one, often used to draw a stark line between "us" and "them," the "civilized" and the "barbarian." Traditionally, the hallmarks of civilization have been urbanization, agriculture, and writing. These markers, born from a Eurocentric perspective, have been used to create a hierarchy of societies, with those exhibiting these traits placed at the pinnacle of human development. However, this narrow definition fails to capture the rich tapestry of human societies and the diverse ways in which they have flourished throughout history.

Urbanization: More Than Just Concrete Jungles

While cities have undoubtedly been centers of innovation and cultural exchange, equating urbanization with civilization overlooks the complex social structures and sophisticated knowledge systems of non-urban societies.

Case Study: The Amazonian Indigenous Communities: Deep in the Amazon rainforest, indigenous groups like the Kayapo have developed sustainable ways of life that integrate seamlessly with their environment. Their intricate understanding of the rainforest's ecosystem, their sophisticated social structures, and their rich cultural traditions challenge the notion that cities are the sole bastions of civilization. Recent research using advanced LiDAR technology has revealed evidence of extensive urban planning and sophisticated agricultural practices in pre-Columbian Amazonian societies, further undermining the traditional association of civilization with European-style urban centers.

Agriculture: Beyond the Plow

The development of agriculture has been hailed as a pivotal moment in human history, enabling settled lifestyles and population growth. However, this narrative often ignores the diverse forms of food production and land management practiced by different societies.

Example: Australian Aboriginal Land Management: For over 65,000 years, Aboriginal Australians have practiced sophisticated land management techniques, including controlled burning, to maintain biodiversity and ensure the sustainable use of resources. These practices, often dismissed as "primitive," demonstrate a deep understanding of ecological processes and a long-term perspective on environmental stewardship.

Writing: Beyond the Written Word

Writing has been seen as a key marker of civilization, enabling the transmission of knowledge across generations and the development of complex societies. However, this view neglects the rich oral traditions and other forms of communication that have played a vital role in many cultures.

Example: The Polynesian Navigation System: The Polynesian people, master navigators who colonized vast stretches of the Pacific Ocean, relied on a complex system of oral traditions, celestial navigation, and intricate knowledge of ocean currents and wind patterns. Their ability to navigate across thousands of miles of open ocean without written maps challenges the idea that writing is essential for advanced knowledge and societal development.

The Limitations of a Eurocentric Lens

The traditional markers of civilization are rooted in a Eurocentric worldview that has often been used to justify colonialism and the suppression of indigenous cultures. By prioritizing certain characteristics, such as urbanization and writing, this perspective has marginalized the achievements and complexities of societies that have followed different paths.

Redefining Civilization for the 21st Century

In the 21st century, we need a more inclusive and nuanced understanding of civilization. Rather than relying on a narrow set of criteria, we should recognize the diversity of human societies and the different ways in which they have adapted to their environments and created meaningful lives. This requires a shift away from a Eurocentric perspective and a greater appreciation for the knowledge and traditions of indigenous cultures.

Subtopic 2: Complexity and Interconnectivity: Exploring the role of social networks, trade, and information exchange in shaping civilizations and their resilience.

The Threads That Bind: How Networks Shape Civilizations

The image of a civilization as a monolithic entity, isolated and self-sufficient, is a misconception. In reality, civilizations have always been shaped by intricate networks of interaction, trade, and information exchange.

These connections, like threads in a tapestry, weave together diverse communities and contribute to their resilience in the face of challenges.

Social Networks: The Fabric of Society

Social networks, the intricate web of relationships between individuals and groups, are the backbone of any civilization. These networks facilitate cooperation, knowledge sharing, and collective action, enabling societies to thrive.

Example: The Kula Ring of the Trobriand Islands: This elaborate system of ceremonial exchange, involving the circulation of shell necklaces and armbands across a vast network of islands, demonstrates the power of social networks to foster trust, maintain peace, and facilitate trade. The Kula Ring, studied extensively by anthropologist Bronislaw Malinowski, highlights how seemingly simple exchanges can have profound social and economic implications, strengthening bonds between communities and promoting cultural cohesion.

Trade: The Lifeblood of Civilizations

Trade has been a driving force in human history, connecting distant communities and facilitating the exchange of goods, ideas, and technologies. Trade routes have acted as arteries, carrying not just commodities but also cultural influences, religious beliefs, and innovations that have shaped the course of civilizations.

Case Study: The Silk Road: Spanning thousands of miles, the Silk Road connected the East and the West, facilitating the exchange of silk, spices, and other valuable goods. But the Silk Road was more than just a trade route; it was a conduit for the transmission of Buddhism, Islam, and other religions, as well as scientific knowledge and technological innovations. The Silk Road exemplifies how trade networks can act as vectors for cultural diffusion and societal transformation.

Information Exchange: The Currency of Progress

The flow of information is essential for the growth and resilience of civilizations. The ability to share knowledge, learn from others, and adapt to changing circumstances is crucial for societies to navigate challenges and thrive.

Example: The Islamic Golden Age: During this period of intellectual flourishing, scholars from across the Islamic world gathered in centers like Baghdad and Cordoba, translating ancient Greek texts, developing new mathematical concepts, and making significant advancements in astronomy, medicine, and other fields. The Islamic Golden Age demonstrates the power of information exchange to foster innovation and propel societal progress.

Resilience Through Interconnectivity

Interconnectedness not only fosters growth and innovation but also enhances the resilience of civilizations. By diversifying resources, sharing knowledge, and fostering cooperation, networks can help societies withstand shocks and adapt to changing circumstances.

Case Study: The Inca Road System: The Inca Empire, stretching across the Andes Mountains, developed an extensive road network that facilitated communication, trade, and the movement of armies. This network played a crucial role in the empire's resilience, enabling the efficient distribution of resources and the rapid response to natural disasters and internal rebellions.

The Digital Age: A New Era of Interconnectivity

In the 21st century, the rise of digital technologies has ushered in a new era of interconnectedness. The internet, social media, and other digital platforms have created unprecedented opportunities for communication, collaboration, and information sharing, connecting individuals and communities across the globe. However, this increased connectivity also brings new challenges, such as the spread of misinformation, the erosion of privacy, and the potential for increased social fragmentation.

Navigating the Complexities of the Interconnected World

Understanding the role of networks in shaping civilizations is crucial in the 21st century. As our world becomes increasingly interconnected, we need to foster cooperation, promote the responsible use of technology, and ensure that the benefits of globalization are shared equitably. By recognizing the importance of networks, we can build more resilient and sustainable societies.

Subtopic 3: Cultural Innovation and Adaptation: Analyzing the interplay between technological advancements, cultural practices, and environmental adaptation in the development of civilizations.

The Dance of Progress: How Culture, Technology, and Environment Shape Civilizations

Civilizations are not static entities; they are dynamic systems constantly evolving in response to new challenges and opportunities. This evolution is shaped by a complex interplay between cultural innovation, technological advancements, and environmental adaptation. These three forces, like dancers in an intricate ballet, influence and respond to each other, driving the trajectory of civilizations.

Cultural Innovation: The Spark of Change

Culture, the shared values, beliefs, and practices of a society, provides the framework within which innovation occurs. Cultural values can encourage or discourage certain types of innovation, while cultural practices can provide the inspiration for new technologies and ways of adapting to the environment.

Example: The Invention of the Printing Press: While the technology for printing existed in China for centuries, it was in Europe, with its emphasis on literacy and the dissemination of knowledge, that the printing press truly revolutionized society. The cultural context provided fertile ground for this technological innovation to flourish, leading to the spread of ideas, the rise of literacy, and the transformation of European civilization.

Technological Advancements: Shaping Possibilities

Technological advancements provide new tools and capabilities that can dramatically alter the course of civilizations. From the invention of the wheel to the development of the internet, technology has played a pivotal role in shaping how societies organize themselves, interact with their environment, and solve problems.

Case Study: The Agricultural Revolution: The development of agriculture, a technological innovation that allowed for the domestication of plants and animals, fundamentally transformed human societies. It enabled settled lifestyles, increased food production, and led to the growth of cities and complex social structures. However, it also had profound environmental impacts, leading to deforestation, soil erosion, and the alteration of ecosystems.

Environmental Adaptation: The Necessity of Change

The environment plays a crucial role in shaping the development of civilizations. Societies must adapt to their environment to survive and thrive, developing technologies and cultural practices that allow them to utilize resources, cope with challenges, and maintain a sustainable relationship with their surroundings.

Example: The Inuit People of the Arctic: The Inuit have developed a unique culture and technology that allows them to thrive in one of the harshest environments on Earth. Their knowledge of the Arctic ecosystem, their specialized tools for hunting and fishing, and their warm clothing and housing demonstrate the remarkable capacity of humans to adapt to even the most challenging environments.

The Interplay of Forces: A Dynamic Dance

The relationship between cultural innovation, technological advancements, and environmental adaptation is not linear but rather a dynamic interplay. Technological innovations can drive cultural change, while cultural values can influence the direction of technological development. Environmental challenges can spur both cultural and technological innovation, leading to new ways of adapting and thriving.

Case Study: The Polynesian Voyaging Canoes: The Polynesian people, master navigators who colonized vast stretches of the Pacific Ocean, developed sophisticated voyaging canoes that were both technologically advanced and culturally significant. These canoes, built using traditional knowledge and techniques, allowed them to navigate across vast distances, explore new islands, and establish thriving communities. This example highlights the interplay between technology, culture, and environment in shaping the trajectory of a civilization.

The Anthropocene Challenge: Adapting to a Changing Planet

In the Anthropocene, the current geological epoch in which human activity is the dominant influence on the environment, the interplay between culture, technology, and environment takes on new urgency. As we face the challenges of climate change, resource depletion, and environmental degradation, we need to develop new ways of thinking and acting that prioritize sustainability and resilience.

Redefining Progress in the 21st Century

The 21st century demands a redefinition of progress that goes beyond technological advancement and economic growth. We need to embrace a more holistic understanding of progress that takes into account social well-being, environmental sustainability, and cultural diversity. By fostering cultural innovation, harnessing technology responsibly, and adapting to our changing environment, we can create a more just and sustainable future for all.

Subtopic 4: The Anthropocene Challenge: Reassessing the concept of civilization in the context of the Anthropocene, considering the impact of human activities on Earth's systems and the need for sustainable practices.

The Anthropocene Reckoning: Redefining Civilization for a Planet in Peril

The Anthropocene, the proposed geological epoch defined by humanity's profound impact on Earth's systems, presents a stark challenge to our traditional notions of civilization. No longer can we view ourselves as separate from nature, exerting dominion over a passive planet. Our actions have global consequences, altering the very fabric of the Earth system and threatening the stability of the biosphere. In this context, the concept of civilization must be reassessed, moving beyond anthropocentric perspectives and embracing a more eco-centric worldview that recognizes our interconnectedness with the natural world.

The Unintended Consequences of "Progress"

The hallmarks of traditional civilization - urbanization, agriculture, and industrialization - have come at a steep environmental cost. Deforestation, pollution, climate change, and mass extinction are the unintended consequences of our pursuit of "progress." The Anthropocene forces us to confront the uncomfortable truth that our "civilized" ways of life are undermining the very foundations upon which they are built.

Example: The Plastic Pollution Crisis: The ubiquitous presence of plastic in our oceans, landfills, and even our own bodies is a stark reminder of the unintended consequences of our consumerist culture. Plastic, a marvel of modern technology, has become a symbol of our unsustainable practices, highlighting the disconnect between our technological prowess and our ecological responsibility.

The Limits to Growth: Confronting Planetary Boundaries

The Anthropocene challenges the assumption that human progress can continue indefinitely. We are exceeding planetary boundaries, pushing Earth's systems beyond their capacity to absorb our impacts. Climate change, ocean acidification, biodiversity loss, and resource depletion are all signs that we are living beyond our means. The concept of civilization must now grapple with the limits to growth and the need for a more sustainable relationship with the planet.

Case Study: The Collapse of Easter Island: The tragic fate of Easter Island, where a once-thriving civilization collapsed due to deforestation and resource depletion, serves as a cautionary tale for the Anthropocene. It highlights the dangers of unsustainable practices and the need for societies to live within ecological limits.
Redefining Civilization for the Anthropocene

In the Anthropocene, the concept of civilization must be redefined to encompass ecological responsibility and sustainability. A truly civilized society is one that recognizes its interconnectedness with the natural world and strives to live in harmony with its environment. This requires a shift away from anthropocentric perspectives and the embrace of a more eco-centric worldview that values the health of the planet as much as human well-being.

Key Principles for an Anthropocene Civilization:

Sustainability: Living within ecological limits, ensuring that our consumption of resources does not exceed the planet's capacity to regenerate them.
Resilience: Building societies that can withstand shocks and adapt to changing environmental conditions.
Equity: Ensuring that the benefits of civilization are shared equitably and that future generations inherit a healthy planet.
Interconnectedness: Recognizing the interconnectedness of all living things and fostering a sense of responsibility for the well-being of the planet.

Embracing a New Narrative

The Anthropocene challenges us to rethink our place in the world and to embrace a new narrative of civilization, one that is grounded in ecological responsibility and social justice. This is not a time for despair but a time for action. By redefining civilization for the Anthropocene, we can create a more just and sustainable future for all.

Examples of Anthropocene-Aware Initiatives:

The Transition Town Movement: Grassroots initiatives focused on building community resilience and reducing dependence on fossil fuels.
Regenerative Agriculture: Farming practices that restore soil health, sequester carbon, and enhance biodiversity.
Circular Economy: Economic models that prioritize reuse, recycling, and the elimination of waste.
Indigenous Knowledge Systems: Drawing on the wisdom of indigenous cultures to develop sustainable practices and foster a deeper connection with nature.

The Path Forward

The Anthropocene presents a profound challenge to humanity, but it also offers an opportunity to redefine civilization and create a more just and sustainable future. By embracing ecological responsibility, fostering social equity, and harnessing the power of innovation, we can navigate the challenges of the Anthropocene and build a civilization that thrives in harmony with the planet.

The Unstable Climate: When Civilizations Crumble

Subtopic 1: Reading the Past: Exploring the methods used to reconstruct past climates, including ice core analysis, tree rings, and sediment records.

Imagine holding a piece of ice in your hand that is thousands of years old, or counting the rings of a tree that has stood for centuries. These natural archives hold clues to Earth's past climate, like pages in a history book waiting to be deciphered. Scientists, like detectives, use various methods to reconstruct past climates, piecing together the puzzle of Earth's history.

Ice Cores: Frozen Time Capsules

Ice cores, extracted from glaciers and ice sheets, are like time capsules preserving layers of snow that fell over millennia. Each layer traps tiny bubbles of ancient air, providing a snapshot of the atmosphere's composition at that time. By analyzing the gases, such as carbon dioxide and methane, scientists can track how greenhouse gas concentrations have changed over time.

For example, ice cores from Antarctica have revealed that current levels of carbon dioxide are higher than at any point in the past 800,000 years, highlighting the unprecedented nature of human-induced climate change.

Tree Rings: Nature's Historians

Trees, with their annual growth rings, are nature's historians. Each ring's width reflects the growing conditions of that year. Wider rings indicate favorable conditions, such as ample rainfall and warm temperatures, while narrower rings suggest drought or cold spells.

By studying tree ring patterns from different regions, scientists can reconstruct past climate variations across continents. For instance, tree rings from the American Southwest have revealed a history of megadroughts, far exceeding any droughts experienced in the instrumental record.

Sediment Records: Layers of Time

Sediments, deposited in lakes and oceans, accumulate over time, preserving a record of past environmental conditions. The types of pollen grains, microscopic organisms, and minerals found in these layers can reveal changes in vegetation, temperature, and precipitation.

For example, sediment cores from the Arctic Ocean have shown that the Arctic was much warmer and wetter during the early Holocene period, about 10,000 years ago, than it is today.

Paleoclimatology: The Science of Past Climates

These methods, along with others like coral reefs and cave formations, form the foundation of paleoclimatology, the study of past climates. By understanding how climate has changed in the past, scientists can better predict future changes and assess the potential impacts on human societies and ecosystems.

Climate Proxies: Indirect Indicators

These natural archives are called climate proxies, as they provide indirect indicators of past climate conditions. They are like pieces of a puzzle, each contributing to a more complete picture of Earth's climate history.

Historical Climate Data: Written Records

In addition to natural archives, historical records, such as diaries, ship logs, and crop records, can provide valuable insights into past climates, especially over the past few centuries. These records can help fill in the gaps where natural archives are scarce or incomplete.

The Importance of Understanding Past Climates

Understanding past climates is not just an academic exercise; it has profound implications for our future. By studying how climate has changed in the past, we can better understand the natural variability of the climate system and the potential impacts of human activities. This knowledge can help us make informed decisions about how to adapt to a changing climate and mitigate its effects..

Subtopic 2: The Fall of Empires: Analyzing case studies of civilizations that collapsed due to climate change, such as the Maya, the Akkadians, and the Classic Maya.

History is littered with the remnants of once-great civilizations that crumbled into dust. While many factors contribute to societal collapse, climate change has played a significant role in the downfall of many empires. Let's delve into the stories of three such civilizations – the Akkadians, the Classic Maya, and the Tiwanaku – to understand how climate change can destabilize even the most powerful societies.

The Akkadian Empire: A Mesopotamian Tragedy

Around 2300 BCE, in the fertile crescent of Mesopotamia, the Akkadian Empire flourished. Sargon the Great united city-states under his rule, creating a vast empire known for its advancements in writing, mathematics, and law. However, this prosperity was short-lived.

Around 2200 BCE, a devastating drought struck the region. Evidence from marine sediment cores in the Gulf of Oman reveals a sudden shift towards arid conditions. This abrupt climate change led to crop failures, famine, and widespread social unrest. Weakened and vulnerable, the Akkadian Empire succumbed to invasions from neighboring tribes, marking the end of their reign.

The Classic Maya: A Rainforest Riddle

The Classic Maya civilization, known for its magnificent pyramids, intricate hieroglyphic writing, and advanced astronomical knowledge, thrived in the rainforests of Central America from 250 to 900 CE. However, this sophisticated society faced a series of droughts that ultimately led to its collapse.

Paleoclimate data from lake sediments and cave formations reveal a pattern of prolonged droughts during the Terminal Classic period (800-900 CE). These droughts, coupled with deforestation and overpopulation, strained the Maya's resources and led to political instability, warfare, and the abandonment of major cities.

The Tiwanaku: Andean Adaptation and Decline

High in the Andes Mountains of South America, the Tiwanaku civilization flourished from 500 to 1100 CE. They developed innovative agricultural techniques, such as raised fields and irrigation systems, to thrive in the harsh environment. However, even their ingenuity could not withstand the prolonged drought that struck the region around 1000 CE.

Lake sediment records from Lake Titicaca reveal a dramatic drop in water levels, indicating a severe drought. This drought, coupled with social and political unrest, led to the decline of the Tiwanaku civilization.

Lessons from the Past

These case studies highlight the vulnerability of even advanced civilizations to climate change. Drought, famine, and social unrest, triggered by climate shifts, can destabilize societies and lead to their collapse. These ancient stories serve as a stark reminder of the importance of understanding and adapting to climate change in the present day.

Subtopic 3: The Domino Effect: Investigating the complex interplay of factors that contribute to civilizational collapse, including drought, famine, disease, conflict, and social unrest.

Civilizational collapse is rarely a single-factor event. It's more like a domino effect, where one event triggers a cascade of consequences, ultimately leading to the downfall of a society. Climate change often acts as the first domino, setting in motion a chain reaction of interconnected factors.

Drought: The Catalyst

Drought, a prolonged period of abnormally low rainfall, is a common trigger for societal collapse. It disrupts agricultural production, leading to food shortages and famine. This can weaken a society, making it vulnerable to other stressors.

For example, the prolonged drought that contributed to the collapse of the Akkadian Empire led to widespread famine and social unrest, making them susceptible to invasions from neighboring tribes.

Famine: The Hunger Pains

Famine, a severe shortage of food, is a direct consequence of drought and agricultural decline. It can lead to malnutrition, disease, and death, weakening the population and straining social structures.

During the Great Irish Famine (1845-1849), a potato blight, exacerbated by social and political factors, led to widespread famine and the death of an estimated one million people. This devastating event triggered mass emigration and had long-lasting social and economic consequences.

Disease: The Silent Killer

Disease outbreaks often follow in the wake of famine and social disruption. Malnourished populations are more susceptible to diseases, and overcrowding and poor sanitation can accelerate their spread.

The Black Death, a bubonic plague pandemic that struck Europe and Asia in the mid-14th century, wiped out an estimated 30-60% of Europe's population. This devastating pandemic had profound social, economic, and cultural impacts, contributing to the decline of feudalism and the rise of the Renaissance.

Conflict: The Struggle for Resources

Competition for dwindling resources, such as food and water, can escalate into conflict and warfare. This can further destabilize a society, disrupt trade routes, and lead to the breakdown of social order.

The Rwandan genocide in 1994, while rooted in ethnic tensions, was also fueled by resource scarcity and environmental degradation. Competition for land and water contributed to social unrest and created a fertile ground for violence.

Social Unrest: The Breaking Point

Social unrest, fueled by famine, disease, and conflict, can erode trust in leadership and lead to social disintegration. This can manifest in protests, riots, and even civil war, further weakening a society's ability to cope with challenges.

The French Revolution (1789-1799) was triggered by a combination of factors, including economic hardship, social inequality, and political oppression. A series of poor harvests and rising food prices contributed to widespread discontent and ultimately led to the overthrow of the monarchy.

Complexity Theory: Understanding Interconnectedness

Complexity theory helps us understand how these factors interact and contribute to societal collapse. It emphasizes the interconnectedness of systems and the potential for small events to trigger cascading effects.

The collapse of the Classic Maya civilization is a prime example of complexity theory in action. A combination of factors, including drought, deforestation, overpopulation, and political instability, interacted in complex ways to bring about the downfall of this once-great civilization.

Lessons for the Present

Understanding the domino effect of civilizational collapse is crucial for building resilience in the face of climate change and other challenges. By recognizing the interconnectedness of social, economic, and environmental systems, we can develop strategies to mitigate risks and adapt to change.

Subtopic 4: Lessons for the Present: Drawing lessons from past collapses to understand the vulnerabilities of modern societies to climate change and develop strategies for adaptation and resilience.

The ghosts of fallen civilizations whisper warnings to the modern world. Their stories, etched in stone and ice, reveal the fragility of human societies in the face of climate change. By learning from their mistakes, we can chart a more sustainable path for the future.

Vulnerabilities of Modern Societies

Modern societies, despite technological advancements, are not immune to the impacts of climate change. In fact, our interconnectedness and reliance on complex systems make us even more vulnerable in some ways.

Interdependence: Global trade and supply chains make us reliant on distant resources and vulnerable to disruptions in other parts of the world. Climate-related events, such as droughts or floods, in one region can have ripple effects across the globe.

Urbanization: The concentration of populations in urban centers increases vulnerability to climate-related disasters, such as heat waves, floods, and storms.
Inequality: Social and economic inequalities exacerbate the impacts of climate change, with marginalized communities often bearing the brunt of its effects.
Environmental Degradation: Deforestation, pollution, and overexploitation of resources weaken ecosystems and reduce their ability to buffer against climate change impacts.
Lessons from the Past

The collapse of past civilizations offers valuable lessons for navigating the challenges of climate change:

Resilience is Key: Societies that are able to adapt and respond to changing conditions are more likely to survive. This requires flexibility, innovation, and the ability to learn from past mistakes.
Sustainability Matters: Overexploitation of resources and environmental degradation can undermine a society's long-term viability. Sustainable practices, such as renewable energy and responsible resource management, are essential for building resilience.
Social Cohesion is Crucial: Strong social bonds and trust in leadership can help societies weather crises. Inequality and social unrest can exacerbate the impacts of climate change and lead to instability.
Early Warning Systems are Vital: Investing in early warning systems and disaster preparedness can help reduce the impacts of climate-related disasters. This includes monitoring climate trends, developing emergency response plans, and educating the public about risks.
Strategies for Adaptation and Resilience

Drawing on these lessons, we can develop strategies to build resilience in the face of climate change:

Invest in Sustainable Infrastructure: This includes renewable energy, water conservation, and climate-resilient infrastructure, such as flood defenses and drought-resistant crops.
Promote Social Equity: Addressing social and economic inequalities can strengthen social cohesion and reduce vulnerability to climate change impacts.
Enhance Environmental Stewardship: Protecting and restoring ecosystems can help mitigate climate change and enhance resilience to its effects.
Foster International Cooperation: Climate change is a global challenge that requires international cooperation. Sharing knowledge, technology, and resources can help build resilience across borders.
Embrace Innovation: Technological advancements can play a crucial role in adapting to climate change. This includes developing new crop varieties, improving water management, and creating sustainable energy solutions.
A Call to Action

The stories of past civilizations serve as a powerful reminder that climate change is not just an environmental issue; it's a societal issue. By learning from the past, embracing innovation, and working together, we can build a more resilient and sustainable future for all.

Ghosts of Civilizations Past: Unveiling the Hidden Histories

Subtopic 1: Peering Beneath the Surface: Exploring the use of remote sensing technologies (LiDAR, satellite imagery, ground-penetrating radar) to identify potential archaeological sites and uncover buried landscapes.

Introduction:

Archaeology, the study of human history and prehistory through the excavation of sites and the analysis of artifacts and other physical remains, has traditionally relied on laborious and time-consuming fieldwork. However, with the advent of remote sensing technologies, archaeologists are now able to peer beneath the surface and uncover hidden landscapes without ever lifting a shovel. These technologies, including LiDAR (Light Detection and Ranging), satellite imagery, and ground-penetrating radar (GPR), are revolutionizing the field of archaeology, allowing researchers to identify potential archaeological sites, map buried features, and even reconstruct ancient landscapes.

LiDAR (Light Detection and Ranging)

LiDAR is a remote sensing technology that uses laser pulses to measure distances to the Earth's surface. By emitting laser pulses from an aircraft or drone and measuring the time it takes for the pulses to return, LiDAR can create highly accurate 3D maps of the terrain. In archaeology, LiDAR is particularly useful for identifying subtle topographic features that may indicate the presence of buried archaeological sites. For example, LiDAR has been used to map ancient Maya cities in the dense jungles of

Central America, revealing previously unknown settlements, roads, and agricultural terraces.

Satellite Imagery

Satellite imagery, another powerful remote sensing tool, provides a bird's-eye view of the Earth's surface. By analyzing different wavelengths of light reflected from the ground, archaeologists can identify patterns and anomalies that may indicate the presence of buried archaeological features. For example, satellite imagery has been used to identify ancient Roman roads in Britain, which are often visible as faint linear features in aerial photographs.

Ground-Penetrating Radar (GPR)

GPR is a geophysical method that uses radar pulses to image the subsurface. By emitting radar waves into the ground and measuring the time it takes for the waves to be reflected back, GPR can create detailed profiles of the subsurface, revealing buried walls, foundations, and other archaeological features. GPR is particularly useful for identifying buried structures in urban areas, where excavation is often impractical or impossible.

Case Studies:

Caracol, Belize: LiDAR surveys of the Maya city of Caracol in Belize have revealed a vast urban landscape, including previously unknown residential areas, agricultural terraces, and defensive structures. These discoveries have challenged previous assumptions about the size and complexity of Maya cities.
Angkor Wat, Cambodia: Satellite imagery has been used to map the ancient city of Angkor Wat in Cambodia, revealing a complex network of canals, reservoirs, and

temples. These images have provided new insights into the city's infrastructure and water management systems.
Roman Forum, Italy: GPR surveys of the Roman Forum in Italy have revealed buried structures, including the foundations of temples, shops, and houses. These discoveries have helped archaeologists to reconstruct the Forum's layout and understand its evolution over time.
Conclusion:

Remote sensing technologies are transforming the field of archaeology, providing researchers with powerful tools to identify potential archaeological sites, map buried features, and reconstruct ancient landscapes. As these technologies continue to develop and become more accessible, they are sure to play an even greater role in our understanding of the past.

Subtopic 2: The Art of Excavation: Examining the techniques and methodologies used in archaeological excavations to uncover and interpret the remains of past civilizations.

Introduction:

Archaeological excavation is a meticulous process that involves the systematic uncovering and recording of buried remains. It is an art that requires both scientific rigor and creative interpretation. Archaeologists employ a variety of techniques and methodologies to ensure the careful retrieval and analysis of artifacts, features, and other evidence of past human activity.

Techniques and Methodologies:

Survey and Mapping: Before excavation begins, archaeologists conduct thorough surveys and mapping of the site. This involves using various techniques such as aerial photography, ground-penetrating radar, and magnetometry to identify potential areas of interest. Detailed maps are created to record the location of features and artifacts.

Stratigraphy: Stratigraphy is the study of the layers (strata) of soil and sediment that have accumulated over time. Archaeologists carefully excavate these layers one by one, documenting the changes in soil composition, artifact types, and other features. This allows them to establish a chronological sequence of occupation and understand how the site developed over time.

Grid System: To ensure precise recording of finds, archaeologists often use a grid system to divide the excavation area into smaller units. Each unit is carefully excavated and documented, allowing for the accurate spatial analysis of artifacts and features.

Troweling and Screening: Archaeologists use trowels and other hand tools to carefully remove soil and expose artifacts and features. The excavated soil is then passed through screens to recover smaller artifacts that may have been missed during the initial excavation.

Photography and Documentation: Throughout the excavation process, archaeologists meticulously document their findings through photographs, drawings, and detailed written records. This documentation is crucial for later analysis and interpretation.

Artifact Analysis: Once artifacts are recovered, they are carefully cleaned, cataloged, and analyzed. Archaeologists use a variety of techniques to determine the age, function, and cultural significance of artifacts. This may involve microscopic examination, chemical analysis, and comparison with similar artifacts from other sites.

Examples and Case Studies:

Tutankhamun's Tomb: The excavation of Tutankhamun's tomb in the Valley of the Kings, Egypt, is a classic example of meticulous archaeological excavation. Howard Carter's team carefully documented every step of the process, preserving the tomb's contents for future generations.

Ötzi the Iceman: The discovery of Ötzi the Iceman, a well-preserved mummy from the Copper Age, in the Italian Alps, highlights the importance of careful excavation and artifact analysis. The artifacts found with Ötzi, including his clothing, tools, and weapons, have provided invaluable insights into life in prehistoric Europe.

The Terracotta Army: The excavation of the Terracotta Army in China is an ongoing project that has yielded thousands of life-sized terracotta figures. The careful excavation and documentation of these figures have allowed archaeologists to reconstruct the army's formation and understand its purpose.

Conclusion:

Archaeological excavation is a complex and challenging process that requires a combination of scientific methodology and artistic interpretation. By employing a variety of techniques and methodologies, archaeologists are able to uncover and interpret the remains of past civilizations, providing valuable insights into human history and culture.

Subtopic 3: Lost and Found: Investigating recent discoveries of previously unknown civilizations, such as Göbekli Tepe in Turkey and the city of Mahendraparvata in Cambodia.

Introduction:

Archaeology is a field of constant discovery, and in recent years, researchers have unearthed astonishing evidence of previously unknown civilizations, challenging our understanding of human history and pushing back the timeline of complex societies. These remarkable finds, often hidden in plain sight or buried beneath dense vegetation, are rewriting the narrative of our past.

Göbekli Tepe, Turkey:

Located in southeastern Turkey, Göbekli Tepe is a monumental site that has revolutionized our understanding of the Neolithic period. Dating back to the 10th-8th millennium BCE, it predates Stonehenge and the Egyptian pyramids by thousands of years. The site features massive, T-shaped pillars adorned with intricate carvings of animals and abstract symbols. These pillars are arranged in circular enclosures, suggesting they served a ritualistic purpose.

The discovery of Göbekli Tepe has challenged the traditional view of hunter-gatherer societies as simple and nomadic. The construction of such a complex site, requiring sophisticated engineering and social organization, suggests that these early humans were capable of far greater feats than previously thought. Göbekli Tepe is a testament to the ingenuity and creativity of our ancestors and has opened up new avenues of research into the origins of civilization.

Mahendraparvata, Cambodia:

Hidden deep within the dense jungles of Cambodia, Mahendraparvata is an ancient city that predates the iconic Angkor Wat by centuries. Discovered using LiDAR technology, this lost city was once a thriving center of the Khmer Empire. The LiDAR surveys revealed a complex network of temples, palaces, and roads, indicating a highly organized and sophisticated society.

Mahendraparvata's discovery has shed new light on the early history of the Khmer Empire and its rise to prominence. It also highlights the power of remote sensing technologies in uncovering hidden archaeological treasures. The city's location in a remote and inaccessible area has protected it from looting and development, making it a valuable resource for archaeologists studying the Khmer civilization.

Other Recent Discoveries:

Karahan Tepe, Turkey: Another significant Neolithic site in Turkey, Karahan Tepe, features similar T-shaped pillars to Göbekli Tepe, suggesting a connection between the two sites.
Sacsayhuamán, Peru: This Inca fortress, built with massive stone blocks, continues to puzzle archaeologists with its precise construction and astronomical alignments.
The Antikythera Mechanism: This ancient Greek astronomical calculator, discovered in a shipwreck, demonstrates the advanced technological capabilities of the ancient Greeks.
Conclusion:

These recent discoveries of previously unknown civilizations are revolutionizing our understanding of the past. They demonstrate that human history is far more complex and fascinating than we ever imagined. As archaeologists continue to explore the world, using new technologies and innovative methods, we can expect even more astonishing discoveries in the years to come.

Subtopic 4: Preserving the Past: Discussing the ethical considerations in archaeological research, including the preservation of cultural heritage, repatriation of artifacts, and the impact of tourism and development.

Introduction:

Archaeology is not just about unearthing the past; it's also about preserving it for future generations. Ethical considerations are paramount in archaeological research, guiding how we interact with ancient remains, the communities connected to them, and the impact of our work on the world. This involves navigating complex issues of cultural heritage management, repatriation of artifacts, and the delicate balance between tourism and preservation.

Preservation of Cultural Heritage:

Cultural heritage encompasses tangible remains like sites and artifacts, but also intangible aspects like traditions, languages, and stories passed down through generations. Archaeologists have a responsibility to:

Minimize Damage: Excavation is inherently destructive. Modern archaeology prioritizes minimal intervention, using non-invasive techniques like GPR and LiDAR whenever possible. When excavation is necessary, it's done with meticulous care, documenting everything for future study.
Conserve and Protect: Artifacts are fragile. Proper conservation techniques prevent deterioration, ensuring they can be studied and displayed for years to come. This

includes climate control, specialized storage, and sometimes even advanced scientific analysis to understand material composition and degradation processes.
Site Management: Archaeological sites face threats from looting, natural disasters, and even urban encroachment. Preservation involves protective measures like site monitoring, legal protections, and sometimes controlled access to limit damage from tourism.
Repatriation of Artifacts:

The ownership of artifacts is a complex ethical and legal issue. Many objects now residing in museums were acquired during colonial eras, raising questions of rightful ownership.

Ethical Considerations: Repatriation involves returning artifacts to their communities of origin. This acknowledges the cultural and spiritual significance these objects hold, respecting the continuity of living cultures.
Legal Frameworks: International laws and agreements, like the UNESCO Convention, address the illicit import and export of cultural property. However, repatriation often involves complex negotiations between nations, museums, and indigenous groups.
Case Studies: The repatriation of the Parthenon Marbles from the British Museum to Greece remains a contentious debate. In contrast, the return of Native American remains and sacred objects under NAGPRA in the United States provides a model for addressing historical injustices.
Impact of Tourism and Development:

Archaeological sites attract tourists, providing economic benefits but also posing risks.

Sustainable Tourism: Managing visitor numbers, providing educational resources, and ensuring revenue supports conservation efforts are key. This prevents overcrowding and minimizes damage from foot traffic or souvenir hunting.

Development Pressures: Urban expansion, infrastructure projects, and even agriculture can threaten archaeological sites. Ethical archaeology involves advocating for heritage protection in planning processes, sometimes requiring compromises or even relocation of sites when unavoidable.

Community Engagement: Local communities often have deep connections to archaeological sites. Involving them in decision-making, providing economic opportunities, and respecting their cultural perspectives is crucial for ethical and sustainable site management.

Conclusion:

Preserving the past is an ongoing responsibility. Archaeologists must constantly evaluate the ethical implications of their work, balancing the pursuit of knowledge with the protection of cultural heritage for the benefit of all. This involves collaboration with descendant communities, respecting diverse cultural values, and ensuring that our actions today safeguard the past for future generations.

The Anthropocene Epoch: Humanity's Enduring Mark on Earth

Subtopic 1: A New Geological Age

The Anthropocene, a proposed geological epoch, marks a significant departure from the Holocene, the epoch that began about 11,700 years ago. It signifies the profound and enduring impact of human activities on Earth's systems, leading to irreversible changes that will be visible in the geological record for millennia.

Defining the Anthropocene

The term "Anthropocene" was popularized in the early 2000s by Nobel laureate Paul Crutzen and Eugene Stoermer. It recognizes that human activities have become a dominant force shaping the planet, rivaling natural processes. The Anthropocene Working Group (AWG), an interdisciplinary body of scientists, has been tasked with formally defining the Anthropocene and proposing a suitable geological marker to signify its commencement.

Key Characteristics of the Anthropocene

Several key characteristics distinguish the Anthropocene from previous epochs:

The Great Acceleration: The period after World War II saw an unprecedented surge in human population, economic activity, resource consumption, and technological advancement. This "Great Acceleration" has led to dramatic increases in greenhouse gas emissions, deforestation, urbanization, and pollution, leaving a distinct mark on the planet.

Stratigraphic Markers: Scientists are searching for a clear and globally synchronous marker in the geological record to define the start of the Anthropocene. Potential markers include radioactive fallout from nuclear weapons testing, plastic pollution, and the presence of novel materials like concrete and aluminum.
Global Change Research: The Anthropocene concept has spurred a surge in research across various disciplines, including geology, climatology, ecology, and social sciences. This research seeks to understand the complex interactions between human activities and Earth's systems and to develop solutions for a sustainable future.
Impact on Earth's Systems

The Anthropocene is characterized by significant alterations to Earth's systems, including:

Climate Change: The burning of fossil fuels and other human activities have led to a dramatic increase in greenhouse gas concentrations in the atmosphere, causing global warming and associated climate changes.
Biodiversity Loss: Habitat destruction, pollution, and climate change are driving a mass extinction event, with species disappearing at an alarming rate.
Biogeochemical Cycles: Human activities have disrupted the natural cycles of carbon, nitrogen, phosphorus, and other elements, with far-reaching consequences for ecosystems and human well-being.
Land-Use Change: Deforestation, urbanization, and agriculture have transformed vast areas of the planet, altering landscapes and ecosystems.
The Anthropocene: A New Reality

The Anthropocene represents a fundamental shift in our understanding of the relationship between humans and the planet. It challenges us to recognize our role as a geological force and to take responsibility for the future of Earth. The choices we make today will determine the trajectory of the Anthropocene and the legacy we leave for future generations.

Subtopic 2: The Climate Crisis

The climate crisis is arguably the most pressing challenge of the Anthropocene. The overwhelming scientific consensus confirms that human activities, primarily the burning of fossil fuels, are the dominant cause of observed warming since the mid-20th century. This warming has triggered a cascade of effects across the globe, with profound implications for human societies and natural ecosystems.

Analyzing the Evidence

The evidence for anthropogenic climate change is robust and multifaceted:

Temperature Records: Global average temperatures have risen by approximately 1.1°C since the late 19th century, with the most recent decade (2011-2020) being the warmest on record. This warming trend is consistent across various independent datasets and measurement methods.

Greenhouse Gas Concentrations: Atmospheric concentrations of carbon dioxide, methane, and other greenhouse gases have reached levels unprecedented in at least the last 800,000 years, primarily due to human activities.

Ocean Warming: The oceans have absorbed the vast majority of the excess heat trapped by greenhouse gases, leading to rising ocean temperatures, sea level rise, and ocean acidification.

Melting Ice: Glaciers and ice sheets are melting at an accelerating rate, contributing to sea level rise and disrupting regional water cycles.
Extreme Weather Events: The frequency and intensity of heatwaves, droughts, floods, wildfires, and storms are increasing, with devastating impacts on communities and ecosystems.
Current and Projected Impacts

The impacts of climate change are already being felt worldwide and are projected to intensify in the coming decades:

Sea Level Rise: Coastal communities face inundation, erosion, and saltwater intrusion, displacing millions of people and threatening infrastructure.
Water Scarcity: Changes in precipitation patterns and increased evaporation are exacerbating water scarcity in many regions, leading to conflicts and migration.
Food Security: Crop yields are declining in some regions due to heat stress, drought, and extreme weather events, threatening food security and livelihoods.
Human Health: Climate change is increasing the risk of heat-related illnesses, vector-borne diseases, and respiratory problems, particularly in vulnerable populations.
Ecosystem Disruption: Climate change is altering the distribution and abundance of species, disrupting ecosystems, and increasing the risk of extinction.
Challenges of Mitigation and Adaptation

Addressing the climate crisis requires both mitigation (reducing greenhouse gas emissions) and adaptation (adjusting to the impacts of climate change):

Mitigation: Transitioning to a low-carbon economy requires significant investments in renewable energy, energy efficiency, and sustainable1 transportation. International cooperation and policy frameworks are crucial to achieving ambitious emission reduction targets.

Adaptation: Adaptation measures include building resilient infrastructure, developing drought-resistant crops, and implementing early warning systems for extreme weather events. Equity and justice considerations are paramount in ensuring that vulnerable communities have the resources to adapt.

The Urgency of Action

The climate crisis demands urgent and ambitious action. Delaying action will only increase the costs and risks associated with climate change, jeopardizing the well-being of current and future generations. The choices we make in the coming years will determine the severity of climate impacts and the future of our planet.

Subtopic 3: A World Transformed

Humanity's impact on Earth extends far beyond climate change. Our activities have fundamentally altered the planet's biodiversity, ecosystems, and biogeochemical cycles, leaving a lasting legacy on the planet. This transformation has profound implications for the functioning of Earth's systems and the well-being of all life, including humans.

Consequences for Biodiversity

The Anthropocene is marked by an alarming decline in biodiversity, often referred to as the sixth mass extinction. Human activities are driving species to extinction at a rate far exceeding the natural background rate. The primary drivers of biodiversity loss include:

Habitat Destruction: The conversion of natural habitats to agricultural land, urban areas, and other human-dominated landscapes is the leading cause of biodiversity loss. Deforestation, for example, has decimated vast tracts of tropical rainforest, home to a disproportionate share of the world's species.
Overexploitation: Overfishing, hunting, and poaching are pushing many species towards extinction. The illegal wildlife trade, driven by demand for exotic pets, traditional medicine, and luxury goods, is a significant threat to many iconic species.
Pollution: Pollution from industrial activities, agriculture, and urban areas contaminates air, water, and soil, harming wildlife and disrupting ecosystems. Plastic pollution, in particular, is a growing concern, with vast quantities of plastic waste accumulating in the oceans and harming marine life.
Invasive Species: The introduction of non-native species to new environments can have devastating consequences for native biodiversity. Invasive species can outcompete native species for resources, disrupt food webs, and spread diseases.
Climate Change: Climate change is exacerbating the threats to biodiversity by altering habitats, disrupting migration patterns, and increasing the frequency and intensity of extreme weather events.
Ecosystem Transformations

Human activities have transformed ecosystems across the globe, altering their structure, function, and resilience. Some of the most significant ecosystem transformations include:

Deforestation and Forest Degradation: Forests play a critical role in regulating climate, providing habitat for biodiversity, and supporting human livelihoods. Deforestation and forest degradation, driven by agriculture, logging, and urbanization, have reduced forest cover significantly, with far-reaching consequences for the planet.

Ocean Acidification: The absorption of excess carbon dioxide by the oceans is causing ocean acidification, which threatens marine ecosystems, particularly coral reefs and shellfish.

Freshwater Depletion and Pollution: Human activities are depleting and polluting freshwater resources, threatening aquatic ecosystems and human water security.

Land Degradation and Desertification: Unsustainable agricultural practices, overgrazing, and deforestation are contributing to land degradation and desertification, reducing the productivity of land and increasing the risk of dust storms and other environmental problems.

Disruption of Biogeochemical Cycles

Human activities have disrupted the natural cycles of carbon, nitrogen, phosphorus, and other elements, with cascading effects on ecosystems and human well-being. Some of the most significant disruptions include:

Carbon Cycle: The burning of fossil fuels and deforestation have released vast quantities of carbon dioxide into the atmosphere, disrupting the natural carbon cycle and contributing to climate change.

Nitrogen Cycle: The use of synthetic fertilizers and the burning of fossil fuels have increased the amount of nitrogen in the environment, leading to eutrophication of water bodies, acid rain, and other environmental problems.

Phosphorus Cycle: The mining of phosphorus for fertilizers and other uses is disrupting the natural phosphorus cycle, leading to phosphorus pollution in waterways and potential shortages of this essential nutrient in the future.

Case Studies

The Amazon Rainforest: Deforestation in the Amazon rainforest, driven by agriculture and logging, has led to significant biodiversity loss, habitat fragmentation, and increased carbon emissions.
The Great Barrier Reef: The Great Barrier Reef, the world's largest coral reef system, is threatened by ocean acidification, warming waters, and pollution, leading to coral bleaching and ecosystem decline.
The Aral Sea: The Aral Sea, once one of the world's largest inland seas, has shrunk dramatically due to water diversions for irrigation, leading to desertification, dust storms, and economic hardship.
Navigating the Anthropocene

The challenges of the Anthropocene are immense, but they are not insurmountable. By recognizing the interconnectedness of human and natural systems, we can develop solutions that promote both human well-being and environmental sustainability. This requires a fundamental shift in our values, behaviors, and institutions, as well as technological innovation and international cooperation.

Subtopic 4: Navigating the Anthropocene

The Anthropocene presents humanity with an unprecedented challenge: to navigate a world profoundly altered by our own actions. It demands a fundamental shift in our thinking, from a perspective that views humans as separate from nature to one that recognizes our deep interconnectedness with the Earth's systems. This shift requires not only technological innovation but also social change and ethical considerations.

Pathways for a Sustainable Future

Navigating the Anthropocene requires a multifaceted approach that addresses the complex interplay of environmental, social, and economic factors. Some key pathways for a sustainable future include:

Transitioning to a Circular Economy: Moving away from the linear "take-make-dispose" model towards a circular economy that prioritizes resource efficiency, waste reduction, and reuse. This involves designing products for durability and recyclability, promoting sharing and repair initiatives, and developing innovative technologies for waste management and resource recovery.
Embracing Renewable Energy and Energy Efficiency: Rapidly phasing out fossil fuels and transitioning to renewable energy sources such as solar, wind, and geothermal. This requires significant investments in renewable energy infrastructure, research and development, and policy support. Additionally, improving energy efficiency in buildings, transportation, and industry is crucial for reducing energy demand and greenhouse gas emissions.
Sustainable Food Systems: Transforming food systems to ensure food security and nutrition for a growing population while minimizing environmental impacts. This involves promoting sustainable agriculture practices, reducing food waste, and shifting towards more plant-based diets.
Protecting and Restoring Ecosystems: Conserving and restoring natural ecosystems, such as forests, wetlands, and oceans, to enhance their resilience, biodiversity, and ecosystem services. This includes establishing protected areas, implementing sustainable land management practices, and restoring degraded ecosystems.

Urban Sustainability: Creating sustainable cities that are resilient to climate change, resource-efficient, and provide a high quality of life for their inhabitants. This involves promoting compact and walkable urban design, investing in public transportation, and implementing green infrastructure solutions.

Technological Innovation

Technological innovation plays a crucial role in navigating the Anthropocene. Some key areas of innovation include:

Climate Engineering: Exploring and developing technologies to remove carbon dioxide from the atmosphere or to manage solar radiation, with careful consideration of potential risks and ethical implications.
Biotechnology: Harnessing the power of biotechnology to develop sustainable solutions for agriculture, medicine, and environmental remediation. This includes developing drought-resistant crops, bio-based materials, and bioremediation technologies.
Artificial Intelligence: Utilizing artificial intelligence to optimize resource management, improve energy efficiency, and develop innovative solutions for environmental monitoring and conservation.

Social Change

Navigating the Anthropocene requires profound social change, including:

Shifting Consumption Patterns: Moving towards more sustainable consumption patterns that prioritize quality over quantity, reduce material consumption, and promote sharing and reuse.
Environmental Education and Awareness: Raising public awareness about environmental issues and promoting environmental education to foster a sense of responsibility and empower individuals to make informed choices.
Environmental Justice: Ensuring that the benefits and burdens of environmental policies and actions are

distributed equitably, with particular attention to vulnerable communities.

Ethical Considerations

Ethical considerations are central to navigating the Anthropocene. Some key ethical questions include:

Intergenerational Equity: How can we ensure that future generations inherit a healthy and sustainable planet?
Environmental Justice: How can we address the disproportionate impacts of environmental problems on marginalized communities?
Animal Welfare: How can we protect the welfare of animals in a world increasingly dominated by human activities?
Geoengineering: What are the ethical implications of using technology to intentionally alter Earth's systems?

Future Scenarios

Exploring different future scenarios can help us understand the potential consequences of our choices and guide decision-making. Some possible scenarios include:

Sustainable Development: A scenario in which humanity successfully transitions to a sustainable future, characterized by equitable resource use, environmental protection, and social well-being.
Fortress World: A scenario in which societies become increasingly isolated and unequal, with limited cooperation and a focus on protecting resources for the privileged few.
Breakdown: A scenario in which environmental degradation and social instability lead to widespread conflict and collapse.

Conclusion

Navigating the Anthropocene is the defining challenge of our time. It requires a fundamental shift in our thinking, our values, and our actions. By embracing technological innovation, social change, and ethical considerations, we can create a sustainable future for all. The choices we make today will determine the trajectory of the Anthropocene and the legacy we leave for future generations.

Resilience and Ruin: Lessons from Ancient Societies

Subtopic 1: The Enduring Legacy of Rome: Analyzing the factors that contributed to the longevity of the Roman Empire and the complex interplay of social, economic, and environmental factors that led to its decline.

Introduction

The Roman Empire, a civilization that once dominated the Mediterranean world, left an indelible mark on history. Its longevity, spanning over centuries, is a testament to its resilience and adaptability. However, even the mightiest empires can crumble. The decline of Rome was a complex process, interwoven with social, economic, and environmental factors. This subtopic delves into the factors that contributed to both the rise and fall of this ancient superpower, offering valuable lessons for contemporary societies.

The Rise of Rome: Factors Contributing to Longevity

Military Prowess and Strategic Expansion:
Rome's military strength was a cornerstone of its empire. Its legions, renowned for their discipline and effectiveness, conquered vast territories, bringing wealth and resources back to Rome. The Roman military was not just about brute force; it was also a sophisticated organization capable of adapting to different terrains and enemies. The Romans were skilled in siege warfare, engineering, and logistics, allowing them to conquer and hold onto a vast empire. Moreover, their strategic expansion was not just about conquest; it was also about creating alliances and incorporating conquered peoples into the Roman

system. This approach ensured a degree of stability and cooperation within the empire.

Political and Legal Institutions:
Rome's political system, with its Senate, assemblies, and magistrates, provided a framework for governance and administration. The Roman legal system, codified in the Twelve Tables and later expanded, provided a degree of justice and order. The concept of Roman citizenship, extended to conquered peoples, fostered a sense of belonging and loyalty to the empire. This legal and political framework provided stability and continuity, even during periods of internal strife and leadership changes.

Economic Strength and Infrastructure:
Rome's economy was based on agriculture, trade, and exploitation of natural resources. The empire's vast network of roads, aqueducts, and ports facilitated trade and communication. The Roman economy was also supported by a sophisticated system of taxation and coinage. This economic strength provided the resources necessary to maintain the army, build infrastructure, and support the growing population of the empire.

Social and Cultural Cohesion:
Roman society, though hierarchical, was bound by shared values, traditions, and a common language (Latin). The Romans were also skilled in assimilating conquered cultures, adopting and adapting their gods, customs, and technologies. This cultural cohesion helped to unify the diverse peoples of the empire and create a sense of shared identity.

The Decline of Rome: A Complex Interplay of Factors

Economic Decline and Inequality:

Over time, Rome's economy began to falter. Overreliance on slave labor discouraged technological innovation, while excessive taxation and inflation burdened the population. The gap between the rich and poor widened, leading to social unrest and instability. The decline in trade and the loss of territories further weakened the economy, making it difficult to maintain the army and infrastructure.

Political Instability and Corruption:

The Roman political system, once a source of stability, became increasingly corrupt and dysfunctional. Power struggles, assassinations, and civil wars became common, weakening the empire from within. The emperors, often more concerned with personal power than the welfare of the empire, failed to address the growing problems.

Military Overextension and Barbarian Invasions:

The vastness of the empire made it difficult to defend. The Roman army, once invincible, was increasingly stretched thin and reliant on foreign mercenaries. Barbarian tribes, pushed by migration and climate change, began to encroach on the empire's borders, eventually sacking Rome itself.

Environmental Degradation and Resource Depletion:

Deforestation, soil erosion, and overgrazing contributed to environmental degradation, impacting agricultural productivity. The Roman Empire's reliance on lead for plumbing and other purposes may have also contributed to lead poisoning, affecting public health and cognitive abilities.

Lessons from Rome's Rise and Fall

The Roman Empire's story offers valuable lessons for modern societies:

The importance of sustainable economic practices: Overexploitation of resources and environmental degradation can have long-term consequences.
The need for strong and adaptable institutions: Political stability and a fair legal system are essential for long-term success.
The dangers of inequality and social unrest: A large gap between the rich and poor can lead to instability and conflict.
The importance of investing in infrastructure and technology: A strong infrastructure and technological innovation are crucial for economic growth and resilience.
Conclusion

The Roman Empire's legacy is a reminder that even the most powerful civilizations are not immune to decline. By understanding the factors that contributed to Rome's rise and fall, we can gain insights into the challenges facing our own societies and develop strategies for building a more resilient future.

Subtopic 2: Collapse on Easter Island: Examining the ecological and social dynamics that led to the collapse of the Rapa Nui civilization on Easter Island, highlighting the dangers of resource depletion and environmental degradation.

Introduction

Easter Island, a remote volcanic island in the vast expanse of the Pacific Ocean, holds the remnants of a civilization that met a tragic end. The Rapa Nui people, once thriving, experienced a societal collapse that serves as a stark warning about the perils of unsustainable practices. This subtopic explores the ecological and social dynamics that led to this downfall, emphasizing the interconnectedness of human societies and their environment.

The Rise of Rapa Nui Civilization

The ancestors of the Rapa Nui people arrived on Easter Island around 1200 AD, finding a fertile land covered in lush palm forests. They developed a complex society, known for its monumental stone statues called Moai. These towering figures, carved from volcanic rock, were likely symbols of power, lineage, and religious beliefs. The Rapa Nui people were skilled farmers, cultivating bananas, sweet potatoes, and taro. They also fished in the rich waters surrounding the island.

The Seeds of Collapse: Environmental Degradation

The Rapa Nui's growing population and their practices placed immense pressure on the island's fragile ecosystem. Deforestation, driven by the need for timber to transport and erect Moai, led to soil erosion and loss of fertile land. The disappearance of the palm forests also had cascading effects on the ecosystem, leading to the extinction of native bird species and impacting the island's climate. Overfishing further depleted the island's resources, reducing a crucial food source.

Social Unrest and Conflict

As resources dwindled, competition and conflict intensified within Rapa Nui society. The construction of Moai, once a symbol of unity and prosperity, became a source of rivalry between clans. Evidence suggests that warfare and cannibalism became prevalent as the island descended into chaos. The social fabric that once held the community together began to unravel.

The Collapse

By the 18th century, the Rapa Nui civilization had collapsed. The population had plummeted, and the once-thriving society was reduced to a shadow of its former self. The Moai, abandoned and toppled, stood as silent witnesses to the island's tragic fate. When European explorers arrived in the 18th century, they found a deforested island inhabited by a small, impoverished population.

Lessons from Easter Island

The collapse of the Rapa Nui civilization serves as a cautionary tale:

The Limits to Growth: Even in seemingly isolated environments, resources are finite. Unsustainable practices can lead to ecological collapse and societal decline.
The Interconnectedness of Human and Natural Systems: Human societies are inextricably linked to their environment. Environmental degradation can have profound social and economic consequences.
The Importance of Foresight and Adaptation: Societies must be able to anticipate and adapt to changing environmental conditions to ensure their long-term survival.

Relevance to Modern Society

The story of Easter Island resonates with the challenges facing our world today. Climate change, deforestation, and overexploitation of resources are pushing ecosystems to their limits. The Rapa Nui experience highlights the need for sustainable practices, responsible resource management, and global cooperation to avoid a similar fate.

Conclusion

The collapse of the Rapa Nui civilization on Easter Island is a poignant example of the fragility of human societies in the face of environmental degradation. By understanding the factors that led to their downfall, we can learn valuable lessons about the importance of sustainable practices and the interconnectedness of human and natural systems.

Subtopic 3: Egypt's Dance with the Nile: Investigating how ancient Egyptian civilization adapted to the cyclical floods of the Nile River, demonstrating the importance of understanding and managing natural resources.

Introduction

Ancient Egypt, a civilization renowned for its grandeur and longevity, flourished in the arid landscape of North Africa. The key to its success lay in its intricate relationship with the Nile River. The Nile, with its predictable annual floods, provided the lifeblood for this civilization. This subtopic explores how the ancient Egyptians adapted to the river's rhythms, showcasing their ingenuity in harnessing its power and managing its resources.

The Nile: A Source of Life

The Nile River, the longest in the world, flows northward through Egypt, creating a fertile valley in the midst of the desert. The annual floods, caused by monsoon rains in the Ethiopian highlands, brought with them nutrient-rich silt that replenished the soil. This predictable flooding cycle was the foundation of Egyptian agriculture and, by extension, its civilization.

Adaptation and Innovation

The Egyptians developed sophisticated techniques to manage the Nile's floods and harness its resources:

Basin Irrigation: They constructed a network of canals and basins to capture and distribute floodwaters, ensuring that the fertile silt reached their fields. This system allowed them to cultivate crops like wheat, barley, and flax, which formed the staples of their diet.
Nilometers: These measuring devices, placed along the river, helped predict the height of the floodwaters, allowing farmers to prepare their fields and anticipate the abundance of the harvest.
Shaduf: This simple yet effective irrigation tool, consisting of a lever and bucket, allowed farmers to lift water from the Nile to irrigate their crops during the dry season.

The Nile and Egyptian Culture

The Nile was not just a source of sustenance; it was deeply ingrained in Egyptian culture and religion. The Egyptians revered the river as a god, Hapi, who brought fertility and abundance. The annual flooding was celebrated with festivals and rituals, reflecting its importance in their lives. The Nile also influenced Egyptian architecture, with temples and monuments often aligned with the river's flow.

Challenges and Resilience

While the Nile's floods were generally beneficial, they could also be destructive. Excessive flooding could damage crops and settlements, while low floods could lead to drought and famine. The Egyptians developed strategies to mitigate these risks, such as building dykes and storing grain in granaries. Their ability to adapt to the Nile's variability demonstrates their resilience and understanding of their environment.

Lessons from Egypt's Dance with the Nile

The ancient Egyptians' relationship with the Nile offers valuable lessons for modern societies:

The Importance of Water Management: Sustainable water management is crucial for agriculture, economic development, and societal well-being.
Adapting to Natural Cycles: Understanding and adapting to natural cycles, such as floods and droughts, is essential for long-term resilience.
The Interconnectedness of Culture and Environment: Cultural practices and beliefs can play a significant role in shaping how societies interact with their environment.
Conclusion

Ancient Egypt's success was inextricably linked to its ability to understand and manage the Nile River. Their innovative irrigation techniques, their reverence for the river, and their adaptive strategies provide a compelling example of how human societies can thrive by living in harmony with their natural environment.

Subtopic 4: Building a Resilient Future: Drawing lessons from the successes and failures of past societies to develop strategies for building resilience in the face of environmental change, social unrest, and other challenges.

Introduction

The rise and fall of ancient civilizations offer a compelling lens through which to examine the challenges facing our world today. From the enduring legacy of Rome to the collapse on Easter Island and Egypt's dance with the Nile, history provides invaluable lessons about the complex interplay of social, economic, and environmental factors that shape the destiny of societies. This subtopic draws upon these historical insights to explore strategies for building resilience in the face of contemporary challenges.

Lessons from the Past: Key Takeaways for a Resilient Future

Sustainable Resource Management: The collapse of the Rapa Nui civilization on Easter Island serves as a stark reminder of the consequences of unsustainable practices. Overexploitation of resources, deforestation, and environmental degradation can have devastating effects on societies. To build a resilient future, we must prioritize sustainable resource management, ensuring that our consumption patterns do not exceed the Earth's carrying capacity. This includes transitioning to renewable energy sources, promoting sustainable agriculture, and protecting biodiversity.

Strong and Adaptive Institutions: The longevity of the Roman Empire, in part, can be attributed to its strong political and legal institutions. These institutions provided stability, facilitated trade, and ensured a degree of social cohesion. In today's world, strong and adaptive institutions are crucial for navigating complex challenges such as climate change, pandemics, and economic instability. This includes investing in good governance, promoting transparency and accountability, and fostering international cooperation.

Social Equity and Inclusion: The decline of Rome was exacerbated by growing social inequality and unrest. A large gap between the rich and poor can lead to social instability and conflict, undermining societal resilience. To build a more resilient future, we must prioritize social equity and inclusion, ensuring that all members of society have access to opportunities and resources. This includes investing in education, healthcare, and social safety nets.

Technological Innovation and Adaptation: The ancient Egyptians' ability to harness the power of the Nile through innovative irrigation techniques demonstrates the importance of technological adaptation. In the face of climate change and other challenges, technological innovation is crucial for developing sustainable solutions. This includes investing in research and development, promoting clean technologies, and fostering a culture of innovation.

Understanding and Adapting to Natural Cycles: The Egyptians' reverence for the Nile and their understanding of its cyclical floods highlight the importance of adapting to natural cycles. Climate change is disrupting weather patterns and increasing the frequency and intensity of extreme events. To build resilience, we must invest in climate adaptation measures, such as flood control infrastructure, drought-resistant crops, and early warning systems.

Building a Resilient Future: A Multi-faceted Approach

Building a resilient future requires a multi-faceted approach that addresses the interconnected challenges facing our world. This includes:

Strengthening Global Cooperation: Climate change, pandemics, and other global challenges require international cooperation and collective action. This includes strengthening multilateral institutions, promoting global solidarity, and fostering shared responsibility.

Empowering Communities: Building resilience requires empowering communities to participate in decision-making processes and develop locally appropriate solutions. This includes supporting community-based initiatives, promoting local knowledge, and fostering social cohesion.

Investing in Education and Awareness: Education and awareness are crucial for building a resilient future. This includes promoting environmental literacy, fostering critical thinking skills, and encouraging responsible citizenship.

Conclusion

The successes and failures of past societies offer valuable lessons for building a resilient future. By learning from history, embracing sustainable practices, investing in strong institutions, and fostering social equity, we can create a more resilient and sustainable world for generations to come. The challenges are significant, but by drawing upon the wisdom of the past and embracing innovation, we can navigate the complexities of the 21st century and build a future that is both prosperous and resilient.

The Threads of Culture: Weaving Knowledge Across Generations

Subtopic 1: Oral Histories and Collective Memory: Exploring the role of oral traditions, storytelling, and rituals in preserving and transmitting knowledge in pre-literate societies.

In the vast expanse of human history, before the advent of writing, knowledge was a vibrant, living entity, woven into the very fabric of society through the intricate threads of oral traditions, storytelling, and rituals. These elements, intertwined and inseparable, served as the primary vessels for preserving and transmitting knowledge across generations, ensuring the continuity of cultural heritage and the accumulation of wisdom.

Oral Traditions: The Living Library of Knowledge

Oral traditions, encompassing a rich tapestry of spoken narratives, songs, proverbs, and genealogies, functioned as a dynamic repository of knowledge, encapsulating the history, beliefs, values, and practical skills of a community. Passed down through countless generations, these oral traditions served as a living library, accessible to all members of society, regardless of their social status or literacy.

In many pre-literate societies, elders held a revered position as the keepers of knowledge, their memories etched with the wisdom of their ancestors. They were the living embodiment of history, their words carrying the weight of generations past. Through their storytelling, they imparted not only factual information but also the moral and ethical frameworks that guided their communities.

Storytelling: Weaving Knowledge into Narratives

Storytelling, a universal human practice, played a pivotal role in the transmission of knowledge, transforming abstract concepts and historical events into captivating narratives that resonated with listeners of all ages. These stories, often imbued with symbolism and allegory, served as mnemonic devices, aiding in the retention and recall of information.

Moreover, storytelling provided a platform for exploring complex social and moral issues, allowing communities to grapple with questions of identity, justice, and the human condition. Through the vicarious experiences of characters in stories, individuals could learn from the successes and failures of others, gaining valuable insights into human behavior and the consequences of their actions.

Rituals: Embodied Knowledge and Social Cohesion

Rituals, often intertwined with storytelling and oral traditions, provided a powerful means of transmitting knowledge through embodied experiences. These ceremonies, performed with precision and reverence, reinforced social bonds, instilled cultural values, and imparted practical skills.

Initiation rites, for example, marked the transition from childhood to adulthood, imparting crucial knowledge about societal roles and responsibilities. Agricultural rituals, performed in harmony with the cycles of nature, transmitted essential knowledge about farming practices, ensuring the survival of the community.

Case Studies: The Enduring Power of Oral Traditions

The Aboriginal peoples of Australia provide a compelling example of the enduring power of oral traditions. Their complex songlines, passed down through countless generations, map the landscape, recounting the journeys of ancestral beings and encoding vital ecological knowledge. These songlines serve as a living testament to the profound connection between oral traditions and the natural world.

In the Amazon rainforest, the indigenous peoples possess a vast oral library of knowledge about medicinal plants, accumulated over centuries of observation and experimentation. This knowledge, passed down through generations of healers, provides a vital source of healthcare for these communities, demonstrating the practical applications of oral traditions.

Conclusion: The Threads of Oral Tradition

Oral traditions, storytelling, and rituals, interwoven into the fabric of pre-literate societies, served as the primary threads for weaving knowledge across generations. These dynamic and interactive forms of knowledge transmission ensured the continuity of cultural heritage, fostered social cohesion, and provided a framework for understanding the world. Even in today's digital age, the echoes of these ancient traditions can still be heard, reminding us of the enduring power of the spoken word and the importance of preserving the rich tapestry of human knowledge.

Subtopic 2: The Power of the Written Word: Analyzing the impact of the invention of writing on the development of civilizations, including the spread of information, the codification of laws, and the emergence of complex thought.

The invention of writing, a monumental leap in human history, marked a profound shift in the way knowledge was preserved, transmitted, and accessed. This revolutionary technology, emerging independently in various parts of the world, transformed civilizations, enabling the spread of information across vast distances, the codification of laws for societal order, and the emergence of complex thought.

From Oral to Written: A Cognitive Revolution

The transition from oral to written communication marked a cognitive revolution, altering the very structure of human thought. While oral traditions relied on memory and mnemonic devices, writing provided an external memory system, freeing the mind from the constraints of limited recall. This newfound cognitive freedom allowed for the development of more complex and abstract thought, paving the way for advancements in philosophy, science, and mathematics.

Writing also facilitated the standardization of language, creating a shared system of communication that transcended geographical boundaries and temporal limitations. This standardization enabled the dissemination of knowledge across vast distances and across generations, fostering cultural exchange and intellectual growth.

The Spread of Information: Breaking the Barriers of Time and Space

With the advent of writing, information was no longer confined to the realm of the spoken word, limited by the reach of the human voice and the constraints of memory. Written texts could be copied and distributed, carrying knowledge across vast distances and across generations. This newfound ability to disseminate information fueled the growth of trade networks, facilitated the spread of religious and philosophical ideas, and enabled the exchange of scientific and technological innovations.

The invention of the printing press in the 15th century further amplified the power of the written word, democratizing access to information and accelerating the dissemination of knowledge. Books, once a luxury reserved for the elite, became widely available, empowering individuals with the ability to learn and explore new ideas.

Codification of Laws: Establishing Order and Justice

Writing played a crucial role in the development of legal systems, providing a means to codify laws and establish societal order. Written laws, unlike oral traditions, were less susceptible to misinterpretation and manipulation, ensuring greater transparency and accountability.

The Code of Hammurabi, one of the earliest examples of codified law, etched in stone for all to see, established a system of justice based on the principle of "an eye for an eye, a tooth for a tooth." This code, though harsh by modern standards, provided a framework for resolving disputes and maintaining social order in ancient Mesopotamia.

The Roman legal system, another landmark in the history of law, relied heavily on written documentation, establishing a complex system of courts, judges, and legal procedures. Roman law, with its emphasis on written contracts and legal precedents, has had a profound and lasting influence on legal systems around the world.

Emergence of Complex Thought: Philosophy, Science, and Mathematics

The invention of writing provided a fertile ground for the development of complex thought, enabling the exploration of abstract concepts and the formulation of intricate theories. Philosophers, no longer constrained by the limitations of oral discourse, could record their ideas, analyze arguments, and engage in nuanced debates.

The works of Plato and Aristotle, preserved in written form, laid the foundation for Western philosophy, influencing generations of thinkers. Similarly, the scientific treatises of ancient Greek scholars, such as Euclid's "Elements" and Ptolemy's "Almagest," provided a framework for understanding the natural world.

Mathematics, too, flourished with the advent of writing, as complex calculations and geometric proofs could be recorded and shared. The development of algebraic notation, facilitated by writing, revolutionized mathematics, enabling the expression of abstract concepts and the solution of complex equations.

Case Studies: The Transformative Power of Writing

The ancient Egyptians, with their hieroglyphic writing system, developed a complex civilization, constructing monumental pyramids, temples, and tombs. Their written records, preserved on papyrus and stone, provide invaluable insights into their religious beliefs, social structures, and scientific knowledge.

The Maya civilization of Mesoamerica, with their intricate system of hieroglyphic writing, developed a sophisticated understanding of astronomy, mathematics, and calendar systems. Their written records, carved on stone monuments and painted on codices, document their history, rituals, and scientific achievements.

Conclusion: The Enduring Legacy of the Written Word

The invention of writing, a watershed moment in human history, transformed civilizations, enabling the spread of information, the codification of laws, and the emergence of complex thought. This revolutionary technology, by providing an external memory system and a standardized form of communication, unleashed the power of human intellect, paving the way for advancements in philosophy, science, and mathematics. The enduring legacy of the written word continues to shape our world today, reminding us of the transformative power of this remarkable invention.

Subtopic 3: Institutions of Knowledge: Investigating the role of libraries, archives, universities, and other institutions in preserving, organizing, and disseminating knowledge throughout history.

Throughout history, the preservation and transmission of knowledge has been inextricably linked to the development of institutions dedicated to this noble pursuit. Libraries, archives, universities, and other institutions have emerged as guardians of knowledge, playing a pivotal role in collecting, organizing, and disseminating information, fostering intellectual growth, and shaping the course of civilizations.

Libraries: The Sanctuaries of Knowledge

Libraries, repositories of written knowledge, have served as beacons of learning for millennia. From the ancient clay tablets of Mesopotamia to the modern digital archives, libraries have evolved to accommodate the ever-changing forms of information, but their core mission has remained constant: to provide access to knowledge for all.

The Great Library of Alexandria, established in the 3rd century BC, stands as a testament to the ambition and intellectual curiosity of the ancient world. This magnificent institution, housing an estimated half a million scrolls, attracted scholars from across the Mediterranean, fostering a vibrant intellectual exchange. Though tragically destroyed, its legacy continues to inspire, reminding us of the transformative power of libraries.

In the medieval era, monastic libraries served as centers of learning, preserving classical texts and fostering the development of Christian scholarship. The invention of the printing press in the 15th century revolutionized the production of books, leading to the proliferation of libraries across Europe. These libraries, often housed in universities and private collections, became hubs of intellectual activity, fueling the Renaissance and the Scientific Revolution.

Today, libraries continue to play a vital role in communities around the world, providing access to information in various formats, from printed books and journals to digital databases and online resources. They serve as community centers, offering educational programs, fostering literacy, and promoting lifelong learning.

Archives: Preserving the Records of the Past

Archives, dedicated to the preservation of historical records, provide a window into the past, allowing us to trace the evolution of societies, governments, and individuals. These institutions safeguard a vast array of documents, including government records, personal letters, photographs, maps, and audio-visual materials.

The National Archives of the United States, established in 1934, houses a treasure trove of historical documents, including the Declaration of Independence, the Constitution, and the Bill of Rights. These documents, carefully preserved and made accessible to the public, serve as a testament to the enduring principles of American democracy.

Archives play a crucial role in historical research, providing primary sources that allow scholars to reconstruct the past, challenge existing narratives, and gain new insights into historical events. They also serve as a vital resource for genealogists, legal professionals, and anyone seeking to understand their personal or family history.

Universities: The Crucibles of Knowledge Creation

Universities, institutions of higher learning, have been at the forefront of knowledge creation and dissemination for centuries. These institutions, bringing together scholars and students from diverse backgrounds, foster intellectual curiosity, critical thinking, and the pursuit of knowledge.

The University of Bologna, founded in 1088, is considered the oldest university in the world. This pioneering institution, with its focus on law and the liberal arts, established a model for universities across Europe. The medieval universities, with their emphasis on scholarly debate and the pursuit of truth, played a crucial role in the development of Western thought.

Today, universities continue to be centers of innovation and research, pushing the boundaries of knowledge in fields ranging from science and technology to the humanities and social sciences. They serve as incubators for new ideas, fostering collaboration and providing a platform for the exchange of knowledge.

Museums: Showcasing the Treasures of Knowledge

Museums, institutions dedicated to the preservation and display of artifacts and specimens, provide a tangible link to the past, showcasing the cultural heritage of societies around the world. These institutions house a vast array of objects, from ancient sculptures and paintings to scientific specimens and technological innovations.

The British Museum, established in 1753, houses a vast collection of artifacts from around the world, including the Rosetta Stone, the Parthenon sculptures, and the Egyptian mummies. 1 These objects, carefully preserved and displayed, provide a glimpse into the rich tapestry of human history and cultural diversity.

Museums play a vital role in education and public engagement, offering exhibitions, educational programs, and interactive displays that bring history to life. They serve as cultural centers, fostering appreciation for the arts, sciences, and the diversity of human experience.

Case Studies: The Impact of Institutions of Knowledge

The Renaissance, a period of cultural and intellectual rebirth in Europe, was fueled by the rediscovery of classical texts preserved in monastic libraries and the establishment of new universities. These institutions, by providing access to knowledge and fostering intellectual inquiry, ignited a period of artistic, scientific, and philosophical innovation.

The Scientific Revolution, a period of unprecedented scientific discovery in the 17th century, was driven by the exchange of ideas among scientists and the publication of their findings in scholarly journals. Institutions such as the Royal Society of London, founded in 1660, played a crucial role in promoting scientific inquiry and disseminating knowledge.

Conclusion: The Pillars of Knowledge Preservation

Libraries, archives, universities, and museums stand as pillars of knowledge preservation, playing a vital role in collecting, organizing, and disseminating information throughout history. These institutions, by providing access to knowledge and fostering intellectual inquiry, have shaped the course of civilizations, fueling cultural and scientific advancements. As we navigate the digital age, these institutions continue to adapt and evolve, ensuring that knowledge remains accessible to all and that the legacy of human intellectual achievement is preserved for future generations.

Subtopic 4: The Digital Age: Examining the challenges and opportunities of the digital age for the preservation and transmission of knowledge, considering the impact of the internet, social media, and artificial intelligence.

The digital age, ushered in by the advent of the internet and the proliferation of digital technologies, has fundamentally transformed the landscape of knowledge preservation and transmission. This era of unprecedented connectivity and information accessibility presents both exciting opportunities and daunting challenges for safeguarding and sharing knowledge in the 21st century and beyond.

The Internet: A Global Network of Knowledge

The internet, a vast interconnected network of computers, has revolutionized the way we access and share information. This global platform has democratized knowledge, breaking down geographical barriers and providing access to a wealth of information with unprecedented ease.

Online encyclopedias like Wikipedia, with their collaborative editing model, offer a dynamic and constantly evolving repository of knowledge, accessible to anyone with an internet connection. Digital libraries, such as Project Gutenberg and the Internet Archive, provide access to millions of digitized books, journals, and other media, preserving cultural heritage and making it freely available to the world.

The internet has also fostered the growth of online learning platforms, offering educational opportunities to individuals around the globe. Massive Open Online Courses (MOOCs), offered by prestigious universities and institutions, provide access to high-quality education at little or no cost, democratizing learning and empowering individuals to pursue their intellectual passions.

Social Media: Connecting and Sharing Knowledge

Social media platforms, such as Facebook, Twitter, and Instagram, have emerged as powerful tools for connecting people and sharing information. These platforms facilitate the rapid dissemination of news, ideas, and cultural trends, fostering global conversations and enabling individuals to engage with a diverse range of perspectives.

Social media has also become a platform for citizen journalism, empowering individuals to document events, share their experiences, and challenge official narratives. This democratization of information has the potential to hold power to account and promote transparency.

However, the rapid spread of information on social media also presents challenges. The proliferation of misinformation, disinformation, and "fake news" poses a threat to the integrity of knowledge, requiring critical thinking skills and media literacy to navigate the digital landscape effectively.

Artificial Intelligence: Augmenting Human Intellect

Artificial intelligence (AI), the development of computer systems capable of performing tasks that typically require human intelligence, 1 has the potential to revolutionize the way we preserve and transmit knowledge. AI-powered tools can assist in the digitization and preservation of cultural heritage, automate the translation of languages, and personalize learning experiences.

AI can also be used to analyze vast amounts of data, identifying patterns and insights that would be impossible for humans to discern. This capability has applications in scientific research, historical analysis, and the development of new technologies.

However, the rise of AI also raises ethical concerns. The potential for bias in AI algorithms, the impact on employment, and the implications for privacy and security require careful consideration and responsible development.

Challenges of the Digital Age

The digital age presents a number of challenges for the preservation and transmission of knowledge:

Digital Preservation: The rapid pace of technological change poses a challenge for the long-term preservation of digital information. Ensuring the accessibility and integrity of digital archives requires ongoing efforts to migrate data to new formats and technologies.
Information Overload: The sheer volume of information available online can be overwhelming, making it difficult to filter reliable sources from misinformation and disinformation. Developing critical thinking skills and media literacy is crucial in navigating the digital landscape.
Digital Divide: The unequal access to technology and digital literacy skills creates a digital divide, exacerbating existing inequalities and limiting opportunities for individuals and communities. Bridging this divide requires investment in infrastructure and education.
Copyright and Intellectual Property: The ease of copying and sharing digital information raises concerns about copyright infringement and the protection of intellectual property. Balancing the need to preserve and share knowledge with the rights of creators is a complex challenge.

Opportunities of the Digital Age

The digital age also offers unprecedented opportunities for the preservation and transmission of knowledge:

Global Access to Information: The internet provides access to a wealth of information from around the world, breaking down geographical barriers and democratizing knowledge.
Enhanced Collaboration: Digital technologies facilitate collaboration among researchers, scholars, and students

across disciplines and geographical boundaries, fostering innovation and accelerating the pace of discovery.
Personalized Learning: AI-powered tools can personalize learning experiences, tailoring educational content to individual needs and learning styles.
Preservation of Cultural Heritage: Digital technologies enable the preservation and dissemination of cultural heritage in new and innovative ways, ensuring its accessibility for future generations.
Case Studies: Navigating the Digital Landscape

The digitization of the Dead Sea Scrolls, a collection of ancient Jewish texts discovered in the mid-20th century, provides a compelling example of how digital technologies can preserve and share cultural heritage. The scrolls, fragile and susceptible to damage, have been meticulously digitized, making them accessible to scholars and the public worldwide.

The use of AI in medical research illustrates the potential of this technology to accelerate the pace of discovery. AI algorithms can analyze vast amounts of medical data, identifying patterns and insights that could lead to new treatments and cures for diseases.

Conclusion: Embracing the Digital Future

The digital age presents both challenges and opportunities for the preservation and transmission of knowledge. Navigating this complex landscape requires a critical and thoughtful approach, embracing the potential of digital technologies while addressing the ethical and societal implications. By fostering digital literacy, promoting responsible innovation, and ensuring equitable access to technology, we can harness the power of the digital age to preserve and share knowledge for the betterment of humanity.

Deep Time and the Human Psyche: Confronting the Vastness of Earth's History

Our brains, magnificent as they are, evolved to grasp the timescales of a human lifespan – days, seasons, years, even generations. But deep time, the dizzying expanse of Earth's 4.5 billion-year history, stretches beyond our intuitive grasp like an endless, star-studded night sky. Comprehending such vastness presents a formidable cognitive challenge, one that impacts how we understand both the past and the future.

The Mind's Temporal Toolkit: Built for the Here and Now

Our perception of time is a construct, shaped by our biology and experiences. We rely on internal clocks, rhythmic processes in our brains and bodies, to track the passage of time. These clocks, however, are calibrated to the familiar rhythms of day and night, the waxing and waning of the moon, the cycle of seasons. They falter when confronted with the eons of geological time, where events unfold over millions, even billions, of years.

Think of it like trying to measure the distance to a distant galaxy with a ruler. The tool, while perfectly adequate for everyday tasks, is simply not designed for such immense scales. Similarly, our minds struggle to grapple with the sheer magnitude of deep time. Millions and billions of years become abstract, interchangeable terms, losing their meaning and impact.

Cognitive Biases and Deep Time Blindness

Beyond the limitations of our internal clocks, a host of cognitive biases further cloud our perception of deep time. Here are a few key culprits:

Present Bias: We tend to overvalue the present and discount the future. This makes it difficult to truly appreciate the long-term consequences of our actions, whether it's climate change or the depletion of natural resources. The distant future feels less real, less urgent, than the immediate concerns of our daily lives.
Linear Thinking: We're accustomed to thinking of time as a straight line, progressing steadily from past to present to future. But Earth's history is anything but linear. It's a tapestry of cycles, catastrophes, and slow, incremental changes, interwoven in complex ways. Our linear mindset can make it difficult to grasp the cyclical nature of geological processes and the potential for sudden, dramatic shifts in Earth's systems.
Anthropocentrism: We tend to view the world through a human-centered lens, placing ourselves at the pinnacle of importance. This can lead to a distorted view of deep time, where we see humans as the central actors in Earth's story, overlooking the vast stretches of time before our arrival and the potential for a future beyond our existence.

The Perils of "Deep Time Blindness"

This inability to truly grasp deep time, what we might call "deep time blindness," has profound implications for our understanding of the past and future. It can lead to:

Underestimating the Earth's dynamism: We may perceive Earth's systems as stable and unchanging, failing to appreciate the long history of dramatic transformations – ice ages, volcanic eruptions, mass extinctions – that have shaped our planet.

Misinterpreting the past: We may struggle to comprehend the vast timescales involved in evolutionary processes or geological formations, leading to misunderstandings about the history of life and the Earth itself.
Discounting the future: As mentioned earlier, the distant future can feel abstract and irrelevant, making it difficult to prioritize long-term sustainability and the well-being of future generations.
Case Study: The Yucatan Impact and the Dinosaur's Demise

The Chicxulub impact, the asteroid collision that wiped out the dinosaurs 66 million years ago, provides a stark example of deep time blindness. It's easy to imagine this event as a sudden, catastrophic end to the dinosaurs' reign. But the impact's effects reverberated through deep time, triggering a cascade of environmental changes that unfolded over millennia. The dust and debris ejected into the atmosphere caused a global impact winter, disrupting ecosystems and leading to mass extinctions across the planet. The recovery from this event was a slow, arduous process, spanning millions of years. Comprehending the full scope of the impact requires us to think beyond the immediate devastation and consider the long-term consequences that rippled through deep time.

Overcoming Deep Time Blindness: Tools and Techniques

While our brains may not be naturally equipped to grasp the vastness of deep time, we can develop tools and techniques to expand our temporal horizons. Here are a few examples:

Analogies and metaphors: Relating deep time to more familiar timescales can help make it more concrete. For example, imagining Earth's history compressed into a single year, with humans appearing only in the last few

minutes, can be a powerful way to visualize the vastness of geological time.
Visualizations and simulations: Interactive timelines, animations, and virtual reality experiences can help us explore deep time in a more engaging and intuitive way.
Storytelling and narrative: Weaving narratives around historical events and geological processes can make deep time more relatable and meaningful.
By actively engaging with deep time, we can begin to overcome our cognitive limitations and develop a more profound understanding of our planet's history and our place within it. This expanded perspective is crucial not only for appreciating the wonders of Earth's past but also for making informed decisions about our future.

Subtopic 2: Expanding Our Horizons: Discussing the importance of developing a "deep time" perspective for understanding the interconnectedness of Earth's systems, the long-term consequences of human actions, and our place in the universe.

Research: Big History, Earth system science, astrobiology

Imagine holding a single grain of sand on your fingertip. That grain, seemingly insignificant, holds within its tiny form the echoes of deep time. It might contain fragments of ancient mountains, the skeletal remains of microscopic marine organisms, or even the dust of long-dead stars. This interconnectedness, this weaving together of cosmic and earthly history, is at the heart of a deep time perspective.

Developing this perspective is not merely an academic exercise; it's a fundamental shift in how we understand our place in the universe and the consequences of our actions. It allows us to see beyond the immediate, to recognize the intricate web of relationships that connect us to the past, present, and future of our planet and beyond.

Earth as a System: A Symphony of Interconnected Processes

Earth system science provides a framework for understanding our planet as a complex, interconnected system. It views Earth not as a collection of isolated components but as a dynamic whole, where the atmosphere, oceans, land, and life interact in intricate ways. A deep time perspective allows us to see these interactions unfold over vast stretches of time, revealing patterns and feedback loops that operate on scales far beyond human perception.

Consider the carbon cycle, a fundamental process that regulates Earth's climate. Over millions of years, carbon has moved between the atmosphere, oceans, rocks, and living organisms, maintaining a delicate balance. Human activities, particularly the burning of fossil fuels, have disrupted this ancient cycle, releasing vast amounts of carbon into the atmosphere at an unprecedented rate. A deep time perspective reveals the potential consequences of this disruption, highlighting the risk of triggering feedback loops that could lead to runaway climate change.

The Long Shadow of Human Actions:

Our actions today cast a long shadow into deep time. The plastic we discard, the pollutants we release, the landscapes we alter – these leave lasting imprints on the Earth, impacting not only our own generation but also countless generations to come. A deep time perspective forces us to confront the ethical implications of our choices, recognizing that we are not merely stewards of the present but also inheritors of the past and custodians of the future.

Take, for instance, the issue of nuclear waste disposal. The radioactive materials generated by nuclear power plants remain hazardous for thousands, even millions of years. Finding safe storage solutions requires a deep time perspective, one that considers the long-term stability of geological formations and the potential for future environmental changes.

Our Place in the Universe: A Cosmic Perspective

Deep time extends beyond our planet, encompassing the vastness of the cosmos. Astrobiology, the study of life in the universe, invites us to consider our place within this grand cosmic narrative. By exploring the origins of life on Earth, the conditions required for habitability, and the possibility of life beyond our planet, we gain a deeper appreciation for the rarity and fragility of our own existence.

The recent discovery of exoplanets, planets orbiting other stars, has expanded our understanding of the universe's potential for life. Some of these planets reside within the habitable zone of their stars, where conditions might be suitable for liquid water and, potentially, life as we know it. A deep time perspective allows us to contemplate the vast stretches of time required for life to evolve on these distant worlds, and the possibility that civilizations might rise and fall over cosmic timescales.

Examples and Case Studies:

The Great Oxidation Event: Around 2.4 billion years ago, photosynthetic bacteria began releasing oxygen into Earth's atmosphere, transforming the planet's chemistry

and paving the way for the evolution of complex life. This event, a pivotal moment in Earth's history, highlights the profound impact that life can have on the planet over deep time.
The Anthropocene: Many scientists argue that we have entered a new geological epoch, the Anthropocene, marked by the significant impact of human activities on Earth's systems. This concept underscores the profound influence that humans are having on the planet, shaping its future trajectory for millennia to come.
The Voyager Golden Records: Launched in 1977, the Voyager spacecraft carry golden records containing sounds and images representing life on Earth. These records, intended as messages to potential extraterrestrial civilizations, embody a deep time perspective, recognizing the possibility of communication across vast distances and cosmic timescales.
Cultivating a Deep Time Perspective:

Developing a deep time perspective is an ongoing process, a journey of expanding our horizons and challenging our assumptions. It requires us to:

Embrace complexity: Recognize the interconnectedness of Earth's systems and the intricate web of relationships that shape our planet.
Think long-term: Consider the long-term consequences of our actions, extending our ethical considerations beyond our own generation.
Cultivate humility: Acknowledge our place within the vast expanse of deep time, recognizing the limitations of our knowledge and the fragility of our existence.
By embracing a deep time perspective, we can gain a more profound understanding of our planet, our place in the universe, and the responsibilities we bear as inhabitants of this remarkable world.

Subtopic 3: Art and the Sublime: Analyzing how artists, writers, and musicians have grappled with the concept of deep time, using their creative expressions to evoke a sense of awe, wonder, and humility in the face of Earth's immensity.

Research: Art history, literature, environmental aesthetics

Deep time, with its mind-bending scales and humbling implications, has long served as a muse for artists, writers, and musicians. Unable to fully grasp its immensity through logic or reason alone, we turn to creative expression to explore the emotional and philosophical dimensions of this profound concept. Art, in its various forms, becomes a lens through which we can glimpse the sublime, that overwhelming sense of awe and wonder in the face of something vast and incomprehensible.

Painting the Panorama of Time:

Visual artists have grappled with deep time in myriad ways. Landscape painters, in particular, have often sought to capture the grandeur of Earth's ancient formations, conveying a sense of timelessness and permanence.

The Romantic Sublime: 19th-century Romantic painters like Caspar David Friedrich and J.M.W. Turner often depicted vast, awe-inspiring landscapes, dwarfing human figures and emphasizing the power and immensity of nature. Friedrich's "Wanderer above the Sea of Fog" (1818) exemplifies this approach, with its lone figure perched on a rocky precipice, gazing out at a seemingly infinite expanse of mist and mountains. Such works evoke a sense

of the sublime, highlighting the insignificance of the individual in the face of nature's grandeur.
The American West: Artists like Thomas Moran and Albert Bierstadt played a crucial role in shaping perceptions of the American West in the late 19th century. Their monumental canvases depicted the region's dramatic landscapes – towering mountains, vast canyons, and sweeping plains – conveying a sense of deep time and the forces that shaped this ancient land. Moran's paintings of Yellowstone, for instance, helped to galvanize support for the creation of the first national park, preserving these landscapes for future generations.
Contemporary Approaches: Contemporary artists continue to engage with deep time in innovative ways. Andy Goldsworthy, for example, creates ephemeral sculptures using natural materials like leaves, stones, and ice, documenting their fleeting existence through photography. His works highlight the cyclical nature of time and the constant processes of change and renewal that shape the Earth.
Literary Explorations of Deep Time:

Writers, too, have found fertile ground in the concept of deep time, using language to evoke the vastness of geological history and its impact on human consciousness.

H.P. Lovecraft: Lovecraft's cosmic horror often draws upon the unsettling implications of deep time, confronting characters with ancient, unknowable forces that predate humanity and challenge our understanding of the universe. In "The Call of Cthulhu" (1928), the monstrous Cthulhu, a being from the depths of time and space, embodies the terrifying indifference of the cosmos to human existence.
John McPhee: McPhee's non-fiction works, such as "Basin and Range" (1981) and "Annals of the Former World" (1998), explore geological concepts with clarity and

elegance, making deep time accessible to a wider audience. He weaves together personal narratives, scientific observations, and historical anecdotes, creating a rich tapestry of Earth's history.
Science Fiction: Science fiction often grapples with the implications of deep time, imagining future civilizations, interstellar travel, and the long-term consequences of technological advancements. Olaf Stapledon's "Star Maker" (1937), for instance, explores the vastness of the universe and the cyclical rise and fall of civilizations over cosmic timescales.
Music and the Echoes of Time:

Music, with its ability to evoke emotions and transcend language, offers another powerful medium for exploring deep time.

Classical Music: Composers like Gustav Mahler and Richard Strauss often sought to capture the grandeur and immensity of nature in their works. Mahler's Symphony No. 3 (1896), for example, evokes a sense of the Earth's awakening and the cyclical processes of life and death.
Ambient and Electronic Music: Contemporary artists like Brian Eno and Stars of the Lid create immersive soundscapes that evoke the vastness of space and the slow, unfolding processes of deep time. Their music often features long, drawn-out drones and ethereal textures, creating a sense of timelessness and suspension.
Nature Recordings: The sounds of nature itself – the crashing of waves, the rustling of wind, the calls of birds – can evoke a sense of deep time, reminding us of the ancient rhythms and cycles that govern the Earth.
The Power of Art in the Anthropocene:

In the Anthropocene, the proposed geological epoch defined by human impact on the planet, art plays a crucial role in raising awareness about environmental

issues and fostering a sense of responsibility towards the 1
Earth. Artists, writers, and musicians can use their creative voices to:

Challenge our perceptions: By presenting new perspectives on deep time and the interconnectedness of Earth's systems, they can encourage us to rethink our relationship with the planet.
Inspire action: By evoking a sense of awe, wonder, and connection to the natural world, they can motivate us to protect and preserve Earth's heritage for future generations.
Foster hope: By envisioning alternative futures and exploring the resilience of life, they can offer hope in the face of environmental challenges.
Art, in its many forms, serves as a bridge between the human mind and the vastness of deep time. It allows us to explore the emotional, philosophical, and spiritual dimensions of this profound concept, inspiring awe, wonder, and a renewed sense of responsibility towards our planet.

Subtopic 4: Deep Time Ethics: Exploring the ethical implications of a deep time perspective, considering our responsibilities to future generations, the preservation of Earth's heritage, and the long-term consequences of our actions.

Research: Environmental ethics, future ethics, sustainability

Deep time throws down a profound ethical challenge: How do we act responsibly when our actions reverberate through millennia, impacting generations we will never meet? Traditional ethical frameworks, often focused on immediate consequences and human-centered values, struggle to address the vast temporal scales and complex interconnectedness revealed by deep time. A new ethical perspective is needed, one that expands our circle of concern to encompass future generations, the well-being of the planet, and the preservation of Earth's heritage.

The Rights of Future Generations:

Deep time forces us to acknowledge the existence of future people, individuals who will inherit the consequences of our actions. While we cannot know their specific needs and desires, we can recognize their right to a habitable planet and a fair share of Earth's resources. This intergenerational justice demands that we consider the long-term impacts of our decisions, avoiding actions that might jeopardize the well-being of future generations.

Climate Change: The burning of fossil fuels is a prime example of an action with profound intergenerational consequences. The greenhouse gases we emit today will continue to warm the planet for centuries, leading to rising sea levels, extreme weather events, and disruptions to ecosystems, impacting the lives of countless future generations. A deep time ethic demands that we transition to renewable energy sources and reduce our carbon footprint, ensuring a livable future for those who come after us.
Nuclear Waste: As mentioned earlier, the safe disposal of nuclear waste requires a deep time perspective, considering the long-term stability of storage sites and the potential for future geological changes. Failing to address this issue responsibly could leave a toxic legacy for future generations, jeopardizing their health and well-being.
Resource Depletion: The unsustainable consumption of natural resources, such as minerals, forests, and fisheries, can deplete these resources for future generations. A deep time ethic calls for responsible resource

management, ensuring that future generations have access to the resources they need to thrive.
Preserving Earth's Heritage:

Deep time reveals the Earth as a repository of invaluable heritage, a testament to the long and complex history of life on our planet. This heritage includes not only tangible artifacts and fossils but also the intricate web of life, the diversity of ecosystems, and the geological formations that have shaped our planet over billions of years. A deep time ethic recognizes the intrinsic value of this heritage and calls for its preservation for future generations.

Protecting Biodiversity: The current extinction crisis, driven by human activities, is eroding Earth's biodiversity at an alarming rate. A deep time ethic recognizes the intrinsic value of each species and the importance of preserving the intricate web of life for future generations.
Conserving Geological Wonders: Geological formations, such as mountains, canyons, and caves, provide a window into Earth's deep history, offering insights into the processes that have shaped our planet. A deep time ethic calls for the protection of these natural wonders, ensuring that they remain intact for future generations to appreciate and learn from.
Cultural Heritage: Human history, though a mere blip in the vastness of deep time, is also part of Earth's heritage. Preserving archaeological sites, historical monuments, and cultural traditions allows us to connect with our past and pass on our stories to future generations.
The Long-Term Consequences of Technology:

Technological advancements, while offering immense potential benefits, also carry the risk of unintended consequences that could reverberate through deep time. A deep time ethic calls for careful consideration of the long-term impacts of new technologies, ensuring that they are developed and deployed responsibly.

Artificial Intelligence: The rapid development of artificial intelligence raises ethical questions about the future of humanity and the potential for unintended consequences. A deep time ethic demands that we develop AI in a way that aligns with human values and promotes the long-term well-being of both humanity and the planet.
Geoengineering: Geoengineering technologies, designed to manipulate Earth's systems to mitigate climate change, carry the risk of unintended consequences that could have lasting impacts on the planet. A deep time ethic calls for caution and careful consideration of the long-term risks and benefits of such technologies.
Space Exploration: As we venture further into space, we must consider the potential impact of our actions on other celestial bodies and the long-term implications of space colonization. A deep time ethic demands that we act responsibly in our exploration of the cosmos, respecting the integrity of other worlds and ensuring the sustainability of our ventures.
Cultivating a Deep Time Ethic:

Developing a deep time ethic is an ongoing process, requiring us to expand our moral horizons and challenge our anthropocentric biases. It involves:

Long-term thinking: Considering the long-term consequences of our actions and prioritizing the well-being of future generations.
Holistic perspective: Recognizing the interconnectedness of Earth's systems and the intrinsic value of all life.
Humility: Acknowledging the limitations of our knowledge and the potential for unintended consequences.
Responsibility: Taking responsibility for our actions and their impact on the planet and future generations.

By embracing a deep time ethic, we can navigate the challenges of the Anthropocene and create a more sustainable and just future for all beings, present and future. It is a call to act as responsible stewards of our planet, recognizing that we are part of a grand narrative that extends far beyond our own fleeting existence.

Humanity's Footprint: Reading the Anthropocene in Earth's Archive

Subtopic 1: The First Farmers: Investigating the early impacts of agriculture on Earth's landscapes, using pollen records, soil analysis, and archaeological evidence to trace deforestation, erosion, and changes in biodiversity.

The Dawn of Agriculture: A Profound Shift in Humanity's Relationship with Nature

The transition from nomadic hunter-gatherer societies to settled agricultural communities marked a pivotal moment in human history, one that initiated a cascade of changes that would forever alter the Earth's landscapes and ecosystems. This shift, often referred to as the Neolithic Revolution, began around 10,000 BCE in various regions across the globe, including the Fertile Crescent in the Middle East, the Yangtze River Valley in China, and Mesoamerica.

Paleoecology: Deciphering the Language of Pollen Grains

Paleoecology, the study of past ecosystems, provides a powerful lens through which we can examine the environmental consequences of early agriculture. Pollen grains, microscopic structures produced by plants for reproduction, are remarkably resilient and can be preserved in sediments for thousands of years. By analyzing the types and abundance of pollen grains found in ancient soil layers, scientists can reconstruct past vegetation patterns and infer how humans transformed their surroundings.

Deforestation: A Silent Transformation of the Landscape

One of the most significant impacts of early agriculture was deforestation. As farmers cleared forests to create fields for crops and pastures for livestock, the composition of plant communities underwent dramatic changes. Pollen records from Europe, for example, reveal a decline in tree pollen and an increase in pollen from grasses and weeds associated with cultivated lands. This transformation not only altered the physical landscape but also disrupted the habitats of countless species, leading to changes in biodiversity.

Soil Erosion: The Unseen Toll of Agriculture

Another consequence of early agriculture was soil erosion. The removal of natural vegetation left the soil exposed to the erosive forces of wind and rain, leading to the loss of fertile topsoil and the degradation of land. Archaeological evidence, such as the presence of terraces and other soil conservation measures in ancient agricultural systems, suggests that early farmers were aware of the problem of erosion and attempted to mitigate its effects.

Environmental Archaeology: Uncovering the Traces of the Past

Environmental archaeology, a discipline that combines archaeological methods with ecological principles, provides valuable insights into the interactions between humans and their environment in the past. By excavating ancient settlements, analyzing plant and animal remains, and studying the physical and chemical properties of soils, archaeologists can reconstruct past agricultural practices and assess their impact on the environment.

Agricultural History: Tracing the Evolution of Farming Practices

Agricultural history complements paleoecology and environmental archaeology by providing a historical context for understanding the development of agriculture and its environmental consequences. By studying ancient texts, agricultural treatises, and the evolution of farming tools and techniques, historians can trace the trajectory of agricultural innovation and its impact on the environment.

Case Studies: Illuminating the Past, Informing the Present

The Fertile Crescent: The cradle of agriculture, the Fertile Crescent, offers a compelling case study of the early impacts of agriculture on the environment. Pollen records from the region reveal a gradual decline in forest cover and an increase in pollen from cultivated crops such as wheat and barley. Archaeological evidence also indicates that early farmers in the region practiced irrigation, which led to salinization and the degradation of soils in some areas.

The Amazon Rainforest: Recent research suggests that the Amazon rainforest, often considered a pristine wilderness, was in fact shaped by human activities long before the arrival of Europeans. Archaeological evidence indicates that indigenous communities practiced agriculture in the Amazon for thousands of years, creating patches of cleared land within the forest and influencing the distribution of plant species.

Easter Island: The collapse of the ancient civilization on Easter Island serves as a cautionary tale of the potential consequences of unsustainable agricultural practices. Deforestation, soil erosion, and overexploitation of resources are believed to have contributed to the decline of this once-thriving society.

Conclusion: Lessons from the Past, Challenges for the Future

The transition to agriculture marked a profound shift in humanity's relationship with the environment, one that continues to shape our world today. By studying the early impacts of agriculture on Earth's landscapes, we can gain a deeper understanding of the complex interactions between humans and their environment and learn valuable lessons for the future. As we face the challenges of a growing population and a changing climate, the insights gained from paleoecology, environmental archaeology, and agricultural history can help us develop more sustainable agricultural practices and ensure the long-term health of our planet.

Subtopic 2: The Age of Metals: Analyzing the environmental consequences of ancient mining and metallurgy, including pollution from smelting, deforestation for fuel, and the distribution of heavy metals in sediments and ice cores.

The Dawn of Metallurgy: A Technological Leap Forward, an Environmental Challenge

The discovery and mastery of metals marked a significant turning point in human history, ushering in the Bronze Age and later the Iron Age. These technological advancements brought about profound changes in human societies, enabling the development of new tools, weapons, and infrastructure. However, the extraction and processing of metals also had significant environmental consequences, leaving lasting traces in the Earth's archives.

Archaeometallurgy: Unearthing the Secrets of Ancient Metalworking

Archaeometallurgy, the study of ancient metal production and use, provides valuable insights into the environmental impact of early metallurgy. By analyzing the remains of ancient mines, smelting furnaces, and metal artifacts, researchers can reconstruct the techniques used by ancient metalworkers and assess their environmental footprint.

Pollution from Smelting: A Toxic Legacy

Smelting, the process of extracting metals from ores, releases a variety of pollutants into the environment, including heavy metals such as lead, arsenic, and cadmium. These toxins can contaminate soil, water, and air, posing risks to human health and ecosystems. Archaeometallurgical studies have revealed evidence of heavy metal pollution in the vicinity of ancient smelting sites, sometimes extending for kilometers.

Deforestation for Fuel: Feeding the Furnaces of Industry

The production of metals required vast quantities of fuel, primarily wood and charcoal. This demand led to widespread deforestation in many regions, particularly during the Iron Age when the scale of metal production increased dramatically. The removal of forests not only altered landscapes but also contributed to soil erosion and the loss of biodiversity.

Geochemical Tracing: Unraveling the Story of Metals in the Environment

Geochemical tracing techniques allow scientists to track the movement of metals through the environment, providing clues about the sources and extent of pollution. Sediments and ice cores, which preserve layers of deposited material over time, offer a valuable record of past pollution levels. By analyzing the concentrations of heavy metals in these archives, researchers can reconstruct the history of metal pollution and its impact on the environment.

Environmental History: Contextualizing the Human Impact

Environmental history provides a broader context for understanding the environmental consequences of ancient metallurgy. By examining the social, economic, and political factors that drove the demand for metals, historians can shed light on the complex interplay between human activities and the environment.

Case Studies: Echoes of the Past

Rio Tinto, Spain: The Rio Tinto region in southwestern Spain has been mined for copper, silver, and other metals for over 5,000 years. The intensive mining and smelting activities in this region have left a legacy of heavy metal pollution in the soil, water, and sediments, transforming the landscape and impacting local ecosystems.

The Bronze Age Collapse: The collapse of several Bronze Age civilizations around the 12th century BCE has been linked to a variety of factors, including climate change, social unrest, and environmental degradation. Some researchers have suggested that deforestation and pollution from metal production may have contributed to the decline of these societies.

Ancient China: China has a long history of metal production, dating back to the Bronze Age. Archaeometallurgical studies have revealed evidence of extensive mining and smelting activities in ancient China, as well as the development of sophisticated techniques for metalworking. The environmental impact of these activities is still being investigated, but evidence suggests that pollution from metal production may have been a significant issue in some regions.

Conclusion: The Enduring Legacy of the Age of Metals

The Age of Metals brought about remarkable technological advancements, but it also left a lasting environmental legacy. By studying the consequences of ancient mining and metallurgy, we can gain a deeper understanding of the long-term impacts of human activities on the environment. This knowledge can inform our efforts to develop more sustainable practices for resource extraction and use, ensuring a healthier planet for future generations.

Subtopic 3: The Rise of Cities: Examining the impact of urbanization on Earth's systems, including changes in land use, water resources, and the creation of "urban heat islands" detectable in the geological record.

From Villages to Megacities: The Unfolding Story of Urbanization

The rise of cities represents a defining characteristic of human civilization, a testament to our capacity for social organization, innovation, and collective living. From the first modest settlements in Mesopotamia to the sprawling megacities of the 21st century, urbanization has profoundly reshaped the Earth's landscapes and systems, leaving an indelible mark on the geological record.

Urban Ecology: Deciphering the Complex Dynamics of City Ecosystems

Urban ecology, a relatively new field of study, seeks to understand the ecological processes that operate within urban environments. Cities are unique ecosystems, characterized by high population densities, altered land cover, and modified microclimates. Urban ecologists investigate how these factors influence the distribution and abundance of species, the flow of energy and materials, and the overall functioning of urban ecosystems.

Archaeology of Cities: Unearthing the Past, Informing the Present

The archaeology of cities provides a historical perspective on the development of urban centers and their impact on the environment. By excavating ancient cities, analyzing artifacts, and studying the spatial organization of urban settlements, archaeologists can trace the evolution of urban life and its environmental consequences.

Paleoclimate Studies: Reconstructing the Climate History of Cities

Paleoclimate studies, which investigate past climates, offer valuable insights into the long-term interactions between cities and their environment. By analyzing climate proxies such as tree rings, ice cores, and lake sediments, scientists can reconstruct the climate history of urban regions and assess how urbanization has influenced temperature, precipitation patterns, and other climate variables.

Land Use Change: The Urban Footprint on the Landscape

Urbanization inevitably leads to changes in land use. As cities expand, natural habitats are converted into built environments, altering the composition of plant and animal communities. This transformation can lead to the loss of biodiversity, fragmentation of habitats, and disruption of ecological processes.

Water Resources: The Thirst of Cities

Cities require vast quantities of water to sustain their populations and industries. The demand for water often exceeds the capacity of local water sources, leading to the construction of dams, reservoirs, and other infrastructure to transport water from distant locations. These interventions can have significant environmental impacts, altering river flows, disrupting aquatic ecosystems, and contributing to water scarcity in some regions.

Urban Heat Islands: Cities as Hotspots

Urban areas tend to be warmer than their surrounding rural areas, a phenomenon known as the urban heat island effect. This temperature difference is caused by a combination of factors, including the absorption of solar radiation by dark surfaces, the release of heat from buildings and vehicles, and the reduced evapotranspiration due to the lack of vegetation. Urban heat islands can have significant impacts on human health, energy consumption, and air quality.

The Geological Record: Preserving the Legacy of Cities

The impact of urbanization is not only evident in the present but also preserved in the geological record.

Sediments deposited in lakes, rivers, and coastal areas can contain a wealth of information about past urban activities. For example, the presence of heavy metals, construction debris, and other pollutants in sediments can provide clues about the history of industrialization and urbanization.

Case Studies: Cities Through the Lens of Time

Ancient Rome: The Roman Empire, with its sprawling cities and sophisticated infrastructure, left a lasting imprint on the environment. Archaeological evidence reveals the extensive use of lead pipes for water transport in Roman cities, leading to widespread lead contamination. The construction of roads, aqueducts, and other infrastructure also transformed the landscape, altering river flows and impacting ecosystems.

London: The Industrial Revolution transformed London into a global metropolis, but it also brought about significant environmental challenges. Air pollution from coal burning, sewage disposal in the River Thames, and the expansion of the built environment all contributed to the degradation of the urban environment. However, London also offers an example of how cities can adapt and improve their environmental performance. The implementation of pollution control measures, the creation of green spaces, and the development of sustainable infrastructure have all contributed to making London a more livable and environmentally friendly city.

Modern Megacities: Megacities, with populations exceeding 10 million, face unique environmental challenges. The sheer scale of these urban centers puts immense pressure on resources, generates vast amounts of waste, and contributes to air and water pollution. However, megacities also offer opportunities for innovation and sustainable development. The implementation of green building practices, the development of efficient public transportation systems, and the creation of urban green spaces can all help to mitigate the environmental impact of megacities.

Conclusion: Cities as Agents of Change

Cities are not only centers of human activity but also powerful agents of environmental change. The rise of cities has brought about profound transformations in the Earth's landscapes and systems, leaving lasting traces in the geological record. By understanding the impact of urbanization on the environment, we can develop more sustainable urban planning strategies, mitigate the negative consequences of city living, and create healthier and more resilient urban environments for future generations.

Subtopic 4: A New Stratum: Identifying the distinctive markers of the Anthropocene in the geological record, including plastic pollution, radioactive isotopes from nuclear testing, and the global distribution of novel materials.

The Anthropocene: A New Epoch in Earth's History?

The concept of the Anthropocene, a proposed geological epoch defined by the profound impact of human activities on Earth's systems, has sparked intense debate among scientists and the public alike. While the official recognition of the Anthropocene is still pending, the evidence for its existence is mounting, with researchers identifying a range of distinctive markers in the geological record that distinguish this new epoch from the Holocene, the current geological epoch that began about 11,700 years ago.

Stratigraphy: The Science of Earth's Layers

Stratigraphy, the study of rock layers and their relationships, provides the framework for understanding Earth's history. Geologists use stratigraphic principles to identify and classify different geological units, based on their composition, age, and fossil content. The identification of a new geological epoch, such as the Anthropocene, requires the recognition of distinct stratigraphic markers that differentiate it from previous epochs.

The Anthropocene Working Group: Seeking Scientific Consensus

The Anthropocene Working Group (AWG), a multidisciplinary group of scientists, is tasked with evaluating the evidence for the Anthropocene and making recommendations to the International Commission on Stratigraphy, the body responsible for defining the geological time scale. The AWG is currently assessing various potential markers of the Anthropocene, including those related to human-induced changes in the atmosphere, oceans, and land surface.

Environmental Monitoring: Tracking the Human Impact

Environmental monitoring programs around the world are collecting data on a wide range of environmental parameters, including air and water quality, greenhouse gas concentrations, and biodiversity. This data provides valuable insights into the extent and pace of human-induced changes in the Earth's systems, contributing to the identification of potential markers of the Anthropocene.

Distinctive Markers of the Anthropocene

Plastic Pollution: The proliferation of plastics in the environment is one of the most visible and pervasive signs of the Anthropocene. Plastics are now found in virtually every corner of the planet, from the deepest ocean trenches to the highest mountain peaks. The durability of plastics means that they will persist in the environment for centuries, if not millennia, leaving a lasting legacy in the geological record.

Radioactive Isotopes from Nuclear Testing: The detonation of nuclear weapons during the mid-20th century released a pulse of radioactive isotopes into the atmosphere, which were subsequently deposited across the globe. These isotopes, such as plutonium-239 and cesium-137, provide a clear and unambiguous marker of the Anthropocene in the geological record.

Novel Materials: The Anthropocene is also characterized by the widespread distribution of novel materials, such as concrete, aluminum, and synthetic fibers. These materials, which are not found in nature, have become ubiquitous in the built environment and are increasingly being incorporated into sediments and other geological deposits.

Fossil Fuel Combustion: The burning of fossil fuels, such as coal, oil, and natural gas, has released vast quantities of carbon dioxide and other greenhouse gases into the atmosphere, leading to global warming and other climate changes. The isotopic signature of fossil fuel carbon is distinct from that of natural carbon, providing a clear marker of human influence in the atmosphere and oceans.

Changes in Biodiversity: Human activities have led to significant changes in the distribution and abundance of species, including the extinction of many species and the introduction of invasive species. These changes in biodiversity are reflected in the fossil record and provide further evidence for the Anthropocene.

Case Studies: The Anthropocene in Action

The Great Acceleration: The period since the mid-20th century has been dubbed the "Great Acceleration" due to the rapid increase in human population, economic activity, and resource consumption. This period has seen a dramatic increase in the production and distribution of plastics, the release of greenhouse gases, and the transformation of landscapes.

Ocean Acidification: The absorption of carbon dioxide by the oceans is leading to ocean acidification, which threatens marine ecosystems. The chemical changes associated with ocean acidification are being recorded in the shells of marine organisms, providing a clear marker of human influence in the oceans.

The Plastic Gyres: The accumulation of plastic debris in ocean gyres, such as the Great Pacific Garbage Patch, highlights the scale of plastic pollution and its potential impact on marine life. These plastic accumulations are likely to leave a lasting legacy in the geological record.

Conclusion: A New Chapter in Earth's Story

The Anthropocene represents a new chapter in Earth's history, one in which human activities have become a dominant force shaping the planet's systems. The evidence for the Anthropocene is mounting, with researchers identifying a range of distinctive markers in the geological record. While the official recognition of the Anthropocene is still pending, the concept has already sparked important discussions about the human impact on the environment and the need for sustainable practices. As we continue to shape the planet, it is crucial to recognize our role as geological agents and to take responsibility for the legacy we are leaving for future generations.

Beyond Earth: Searching for Life and Civilizations Among the Stars

Let's start with Subtopic 1: The Drake Equation Revisited.

The Drake Equation, that iconic formula scribbled down by Frank Drake back in 1961, has served as a roadmap for our search for extraterrestrial civilizations. It attempts to quantify the number of detectable civilizations in our galaxy by considering factors like the rate of star formation, the fraction of stars with planets, the number of habitable planets per star, and so on.

But here's the thing: in 1961, we knew next to nothing about exoplanets! We were essentially working with educated guesses. Now, fast forward to 2025, and the game has completely changed. Thanks to missions like Kepler, TESS, and the James Webb Space Telescope, we're swimming in data about exoplanets. We've discovered thousands of these alien worlds, some of them eerily similar to Earth.

This explosion of knowledge allows us to revisit the Drake Equation with fresh eyes and updated parameters. For example:

Rate of star formation: While this factor remains relatively stable, our understanding of star formation in different types of galaxies has become more nuanced.
Fraction of stars with planets: This is where things get exciting! We now know that planets are the rule rather than the exception. Most stars seem to have at least one planet, and many have multiple planetary systems. This dramatically increases the potential number of life-bearing worlds.

Number of habitable planets per star: This is still a bit of a mystery, but we're getting better at identifying planets within the habitable zone, where liquid water could exist on the surface. We're also learning that habitability might depend on factors beyond just the distance from a star, such as the planet's atmosphere and magnetic field.
Fraction of habitable planets where life actually appears: This remains one of the biggest unknowns. We still haven't found definitive proof of life beyond Earth, but ongoing research in astrobiology is shedding light on the conditions under which life might emerge. For instance, the discovery of extremophiles – organisms that thrive in extreme environments on Earth – broadens our understanding of where life could potentially exist.
Fraction of life-bearing planets where intelligent life evolves: This is another big question mark. The evolution of intelligence might be a rare event, or it might be a natural consequence of complex life. We simply don't know yet.
Fraction of civilizations that develop a technology that releases detectable signs of their existence: We're getting more creative in our search for technosignatures, beyond just radio signals. We're looking for signs of advanced technology like Dyson spheres (hypothetical megastructures that encompass a star to capture its energy) or unusual atmospheric compositions that might indicate industrial activity.
Length of time for which such civilizations release detectable signals: This is perhaps the most sobering factor in the equation. Even if intelligent civilizations are common, they might be short-lived. Factors like climate change, resource depletion, or self-destruction could limit the lifespan of a technological civilization.
Reassessing the Drake Equation in 2025 is like assembling a complex puzzle with some pieces finally falling into place. While we still have many unknowns, the picture is becoming clearer. The sheer number of exoplanets, combined with our growing understanding of the

conditions for life, suggests that we might not be alone in the universe.

Subtopic 2: Worlds Beyond Our Own.

It's an incredible time to be alive if you're interested in exoplanets. Thanks to tireless robotic explorers like Kepler and TESS (Transiting Exoplanet Survey Satellite), we've gone from knowing a handful of exoplanets to having a catalog of thousands! And these aren't just any planets; they represent an astonishing diversity of worlds, pushing the boundaries of our imagination.

Think about it:

Hot Jupiters: Gas giants orbiting scorchingly close to their stars, where temperatures soar high enough to melt metal. These were some of the first exoplanets discovered, and they challenged our assumptions about how planetary systems form.
Super-Earths: Rocky planets larger and more massive than Earth, with diverse compositions and potential for unique geological activity. Some might have vast oceans or thick atmospheres, while others could be barren and volcanic.
Mini-Neptunes: Planets with sizes between Earth and Neptune, often shrouded in thick atmospheres. These are incredibly common, but their internal structure and potential for habitability remain mysterious.
Ocean Worlds: Planets or moons covered in deep oceans, possibly harboring life beneath their icy surfaces. Think of Jupiter's moon Europa or Saturn's moon Enceladus, but on a much larger scale.
And of course, we're particularly interested in finding potentially habitable worlds. These are planets that orbit within the "Goldilocks zone" – not too hot, not too cold – where liquid water could exist on the surface. We're finding more and more of these potentially habitable

exoplanets, and some of them are surprisingly close to home!

But finding a planet in the habitable zone is just the first step. To truly understand its potential for life, we need to characterize its atmosphere. This is where spectroscopy comes in. By analyzing the light that passes through a planet's atmosphere, we can identify the presence of different molecules, like water vapor, oxygen, methane, and carbon dioxide. These molecules can act as "biosignatures" – telltale signs of life.

The James Webb Space Telescope, launched in 2021, is a game-changer in this regard. Its powerful instruments allow us to study exoplanet atmospheres in unprecedented detail. Imagine: we can now peer into the skies of distant worlds and search for hints of life!

However, the search for habitable exoplanets and biosignatures is fraught with challenges:

Distance: Even the closest exoplanets are incredibly far away. This makes it extremely difficult to obtain detailed observations.
Faintness: Exoplanets are incredibly faint compared to their host stars. It's like trying to spot a firefly next to a spotlight.
Atmospheric Interference: The atmospheres of exoplanets can be complex and dynamic, making it challenging to interpret spectroscopic data.
False Positives: Some non-biological processes can mimic biosignatures, leading to false detections.
Despite these challenges, the quest to explore worlds beyond our own continues to captivate scientists and the public alike. Each new discovery brings us closer to answering the fundamental question: are we alone?

Subtopic 3: Listening for Signals.

For decades, humanity has been scanning the cosmos, hoping to catch a whisper from another civilization. This is the essence of the Search for Extraterrestrial Intelligence (SETI). While we haven't received a confirmed "hello" yet, the quest continues with renewed vigor and increasingly sophisticated methods.

Radio Astronomy: This classic approach, pioneered by Frank Drake himself, involves using powerful radio telescopes to listen for artificial signals from space. Think of it as trying to pick up a faint radio station amidst the cosmic static.

Targeted Searches: We focus on specific star systems, particularly those with potentially habitable planets, hoping to detect intentional broadcasts or leakage from their technology.
Sky Surveys: We systematically scan large portions of the sky, casting a wide net to capture any unusual signals that might be out there.
The SETI Institute, a non-profit research organization, has been at the forefront of radio SETI for decades. Their Allen Telescope Array, located in California, is a dedicated instrument for SETI research, capable of observing multiple stars simultaneously.

Optical SETI: This newer approach searches for pulsed laser signals, which could be used for interstellar communication. Lasers offer several advantages over radio waves: they can carry more information, and they are less susceptible to interference.

Challenges: Optical SETI is like looking for a needle in a haystack. We need to precisely point our telescopes at

the right place at the right time to catch a fleeting laser pulse.
Technosignatures: Beyond radio waves and lasers, we're also searching for other signs of advanced technology, such as:

Dyson Spheres: These hypothetical megastructures, built to harness the energy of a star, would leave a distinctive infrared signature.
Industrial Pollutants: Advanced civilizations might release pollutants into their atmospheres that could be detectable with our telescopes.
Artificial Satellites: Large-scale space infrastructure, such as orbital rings or space habitats, could be visible as they transit their host star.
Data Analysis: The sheer volume of data generated by SETI observations is staggering. We need sophisticated algorithms and machine learning techniques to sift through this data and identify potential signals. Citizen science projects, like SETI@home, allow volunteers to donate their computer's processing power to help analyze this data.

The search for extraterrestrial intelligence is a long shot, but the potential payoff is immense. Even a single confirmed signal would revolutionize our understanding of the universe and our place within it. And who knows, maybe one day, we'll be the ones sending out a message, hoping to make contact with another civilization across the vast cosmic ocean.

Subtopic 4: The Cosmic Perspective.

The search for extraterrestrial life and civilizations isn't just about scientific curiosity; it's about grappling with profound philosophical questions about our place in the universe. What would it mean for humanity if we discovered that we are not alone? What if, after decades of searching, we find no evidence of other civilizations?

Finding Life: Imagine the headlines: "Scientists Confirm Extraterrestrial Life!" The impact on our society would be seismic.

Redefining Humanity: We would no longer be the sole bearers of intelligence and consciousness in the cosmos. This would challenge our anthropocentric worldview and force us to reconsider our place in the grand scheme of things.
Scientific Revolution: The discovery of extraterrestrial life, especially if it's fundamentally different from life on Earth, would revolutionize biology, chemistry, and our understanding of life's origins.
Cultural Impact: Religion, philosophy, art, and literature would all be profoundly influenced by the knowledge that we are not alone. Imagine the new stories we would tell, the new questions we would ask.
Interstellar Communication: If we discovered an advanced civilization, could we communicate with them? What would we say? What could we learn? The possibilities are both exciting and daunting.
The Great Silence: But what if, despite our best efforts, we find no evidence of other civilizations? This "Great Silence" would raise its own set of profound questions.

Are We Alone? Is life on Earth a cosmic fluke, a rare and precious anomaly in a vast and sterile universe? This possibility is both humbling and terrifying.
The Fermi Paradox: If the universe is teeming with life, as some scientists believe, then where is everybody? This paradox highlights the potential challenges to the long-term survival of civilizations. Perhaps advanced civilizations are inherently self-destructive, or maybe interstellar travel is simply too difficult.

Our Responsibility: If we are truly alone, then the fate of life in the universe rests on our shoulders. We would have a profound responsibility to protect our planet and ensure the survival of our species.
The Long-Term Future: Whether we find extraterrestrial life or not, the search itself forces us to think about the long-term future of civilizations.

Cosmic Evolution: How do civilizations evolve over vast timescales? What are the challenges they face? Are there common patterns in their development?
The Future of Humanity: What is our destiny as a species? Will we become a multi-planetary civilization? Will we eventually venture out into the stars?
Existential Risks: What are the greatest threats to our long-term survival? How can we mitigate these risks and ensure the future of humanity?
The search for extraterrestrial life and civilizations is ultimately a search for ourselves. It's a journey of self-discovery that challenges our assumptions, expands our horizons, and forces us to confront the biggest questions about our existence. Even if we never make contact with another civilization, the quest itself will have transformed us.

Guardians of Deep Time: Ethical Responsibilities in the Anthropocene

Subtopic 1: A Legacy for Future Generations: Exploring the concept of intergenerational justice and the ethical obligations we have to future generations to ensure a habitable planet and a just distribution of resources.

Introduction

The Anthropocene, the current geological epoch in which human activity is the dominant influence on the planet's climate and environment, has brought with it a profound sense of responsibility. We are the first generation to fully understand the extent of our impact on the planet, and we are also the last generation with the opportunity to avert the worst consequences of climate change and environmental degradation. This realization has given rise to the concept of intergenerational justice, which holds that we have an ethical obligation to future generations to leave them a habitable planet and a just distribution of resources.

What is Intergenerational Justice?

Intergenerational justice is a principle of fairness and equity between generations. It recognizes that the actions of the present generation can have significant impacts on the well-being of future generations, and that we have a moral duty to consider those impacts in our decision-making. This principle can be applied to a wide range of issues, including climate change, resource depletion, nuclear waste management, and biodiversity loss.

Ethical Obligations to Future Generations

Our ethical obligations to future generations stem from the recognition that they will inherit the consequences of our actions. We have a duty to ensure that they have the same opportunities to live fulfilling lives as we do. This means ensuring that they have access to clean air and water, a stable climate, and a healthy environment. It also means ensuring that they have access to the resources they need to meet their basic needs and pursue their goals.

Sustainable Development

The concept of sustainable development is closely linked to intergenerational justice. Sustainable development is development that meets the needs of the present without compromising the ability of future generations to meet their own needs. It is about finding a balance between economic growth, social progress, and environmental protection.

Challenges to Intergenerational Justice

There are a number of challenges to achieving intergenerational justice. One challenge is the difficulty of predicting the future. It is difficult to know what the needs and preferences of future generations will be. Another challenge is the problem of collective action. It is difficult to get people to cooperate in the present when the benefits of their actions will not be fully realized for many years to come.

Examples and Case Studies

Climate Change: Climate change is perhaps the most pressing example of an intergenerational justice issue. The

greenhouse gases we emit today will continue to warm the planet for centuries to come, with potentially devastating consequences for future generations.
Nuclear Waste Management: Nuclear waste remains radioactive for thousands of years, posing a significant risk to future generations. Finding a safe and permanent way to dispose of nuclear waste is a major challenge.
Resource Depletion: The depletion of natural resources, such as fossil fuels and minerals, could leave future generations with a lower standard of living.
Biodiversity Loss: The loss of biodiversity is a major threat to the health of the planet and the well-being of future generations.
Conclusion

The concept of intergenerational justice is a powerful one. It reminds us that we are not the only generation that matters, and that we have a responsibility to leave a legacy of sustainability for future generations. By taking action to address climate change, conserve resources, and protect the environment, we can help to ensure that future generations have the opportunity to live fulfilling lives.

Subtopic 2: The Moral Standing of Nature: Examining different ethical perspectives on the value of the natural world, including anthropocentric, biocentric, and ecocentric views, and their implications for environmental decision-making.

Introduction

How we value the natural world profoundly shapes our relationship with it and dictates our actions towards the environment. Do we see nature purely as a resource for human use, or does it possess intrinsic value, worthy of respect and protection regardless of its utility to us? This subtopic delves into the core of environmental ethics, exploring the diverse perspectives on the moral standing of nature and how these perspectives influence environmental decision-making.

Anthropocentrism: Human-Centered Values

Anthropocentrism places humans at the center of the moral universe. From this perspective, the natural world is primarily viewed as a means to fulfill human needs and desires. While anthropocentrism doesn't necessarily condone the reckless exploitation of nature, it prioritizes human interests. Conservation efforts, for example, are often justified based on the benefits they provide to humans, such as clean air, water, and resources for economic activities.

Strengths: Anthropocentrism can be pragmatic, driving action by highlighting the direct impact of environmental issues on human well-being. It can also align with economic principles and incentivize sustainable practices that benefit both people and the planet.
Weaknesses: This perspective can lead to the undervaluation of species and ecosystems that lack obvious human utility. It can also justify environmental degradation in the pursuit of short-term economic gains, potentially overlooking long-term consequences and the intrinsic value of nature.
Biocentrism: Life-Centered Values

Biocentrism extends moral consideration to all living beings. This perspective recognizes the inherent worth of individual organisms, emphasizing their right to exist and flourish. Biocentric ethics advocates for the respectful treatment of animals and plants, challenging practices that cause them harm or suffering.

Strengths: Biocentrism promotes a more compassionate and holistic approach to the environment, recognizing the interconnectedness of all living things. It encourages us to consider the impact of our actions on individual organisms, not just on humans.
Weaknesses: Biocentrism can be challenging to apply in practice, especially when balancing the needs of different species. It can also be difficult to define the boundaries of moral consideration – does it extend to all living things, or only to those with certain capacities, like sentience?
Ecocentrism: Ecosystem-Centered Values

Ecocentrism shifts the focus from individual organisms to the entire ecosystem. This perspective emphasizes the interconnectedness and interdependence of all living and non-living components within an ecological system. Ecocentrism values the health and integrity of ecosystems as a whole, recognizing their intrinsic worth and the vital role they play in sustaining life.

Strengths: Ecocentrism provides a comprehensive framework for environmental decision-making, considering the complex interactions within ecosystems. It promotes a long-term perspective, emphasizing the preservation of ecological processes and the resilience of ecosystems.
Weaknesses: Ecocentrism can be challenging to implement as it may require prioritizing the needs of the ecosystem over the interests of individual organisms or even human communities. Balancing these competing interests requires careful consideration and ethical deliberation.
Implications for Environmental Decision-Making

These ethical perspectives have profound implications for how we approach environmental challenges:

Conservation: Anthropocentrism might support conservation efforts that benefit human populations, while biocentrism might prioritize the protection of endangered species, and ecocentrism might focus on preserving entire ecosystems.
Resource Management: Anthropocentrism might favor resource extraction for human use, while biocentrism and ecocentrism might advocate for more sustainable practices that minimize harm to living organisms and ecosystems.
Climate Change: Addressing climate change requires a multifaceted approach. While anthropocentrism might emphasize technological solutions and economic incentives, biocentrism and ecocentrism might call for more radical changes in human behavior and consumption patterns to reduce our impact on the planet.
Examples and Case Studies

The debate over wolf reintroduction: This highlights the tension between anthropocentric concerns (e.g., livestock losses) and ecocentric values (e.g., restoring ecosystem balance).
The controversy surrounding genetically modified organisms (GMOs): This reflects different perspectives on the moral status of organisms and the potential risks to ecosystems.
International climate negotiations: These illustrate the challenges of balancing national interests (anthropocentrism) with global environmental concerns (ecocentrism).
Conclusion

Understanding the different ethical perspectives on the moral standing of nature is crucial for navigating the complex environmental challenges of the Anthropocene. By considering anthropocentric, biocentric, and ecocentric viewpoints, we can develop more comprehensive and ethically informed solutions that balance human needs with the needs of the planet. Ultimately, our ethical framework will shape the legacy we leave for future generations and determine the fate of the natural world.

Subtopic 3: The Nuclear Legacy: Analyzing the ethical challenges of managing nuclear waste, considering the long-term risks, the need for international cooperation, and the responsibility to future generations.

Introduction

Nuclear technology, while offering a powerful source of energy, leaves behind a legacy that demands careful consideration and ethical responsibility. Nuclear waste, the byproduct of nuclear power generation and other nuclear activities, remains radioactive for extremely long periods, posing significant risks to human health and the environment. Managing this hazardous material presents a complex ethical challenge, requiring us to grapple with long-term risk assessment, the need for international cooperation, and our obligations to future generations.

The Long-Term Risks of Nuclear Waste

The radioactive nature of nuclear waste presents a unique challenge. Unlike other hazardous waste that may degrade over time, nuclear waste remains dangerous for thousands, even hundreds of thousands of years. The risks associated with nuclear waste include:

Radiation Exposure: Exposure to radiation can cause various health problems, including cancer, genetic mutations, and birth defects.
Environmental Contamination: Leaks or accidents at nuclear waste storage facilities can contaminate soil,

water, and air, posing a threat to ecosystems and human populations.
Proliferation Risks: Spent nuclear fuel contains materials that can be reprocessed to create weapons-grade plutonium, raising concerns about nuclear proliferation and security.
The Need for International Cooperation

Nuclear waste management is not just a national issue; it's a global concern. The transboundary nature of radiation means that accidents or mismanagement in one country can have implications for others. International cooperation is essential for:

Sharing Best Practices: Countries with experience in nuclear waste management can share their knowledge and expertise with others, promoting safety and efficiency.
Developing International Standards: Establishing common standards for waste storage, transportation, and disposal can help minimize risks and ensure consistency.
Addressing Proliferation Concerns: International cooperation is crucial for preventing the diversion of nuclear materials for weapons development.
Responsibility to Future Generations

The long-lived nature of nuclear waste makes it a prime example of intergenerational justice. Our decisions today will have consequences for generations to come. We have a moral obligation to:

Minimize Waste Generation: Investing in research and development of nuclear technologies that produce less waste is essential.
Develop Safe Disposal Solutions: Finding safe and permanent disposal solutions for existing waste is a priority. Geological repositories, deep underground storage facilities, are currently considered the most viable option.

Communicate Risks to Future Generations: Developing effective methods to communicate the risks of nuclear waste to future generations is crucial to ensure their safety and well-being.
Ethical Dilemmas and Challenges

Balancing Risks and Benefits: Weighing the benefits of nuclear energy against the risks of long-term waste management is a complex ethical dilemma.
Equity and Justice: Ensuring that the burdens and risks of nuclear waste are distributed fairly across communities and generations is a challenge.
Uncertainty and Prediction: Predicting the long-term behavior of nuclear waste and the potential impacts on future generations involves a degree of uncertainty.
Examples and Case Studies

Yucca Mountain Nuclear Waste Repository (USA): The controversy surrounding this proposed repository highlights the challenges of finding a socially and politically acceptable solution for long-term waste disposal.
The Fukushima Daiichi Nuclear Disaster (Japan): This disaster underscored the potential risks of nuclear accidents and the long-term consequences of radioactive contamination.
International Atomic Energy Agency (IAEA): The IAEA plays a crucial role in promoting international cooperation and developing safety standards for nuclear waste management.
Conclusion

Managing the nuclear legacy is a complex ethical challenge that requires a long-term perspective and a commitment to international cooperation. By recognizing our responsibilities to future generations, we can work towards developing safe and sustainable solutions for nuclear waste management, ensuring a safer and healthier planet for all.

Subtopic 4: Geoengineering Dilemmas: Discussing the ethical implications of geoengineering technologies proposed to mitigate climate change, including solar radiation management and carbon dioxide removal, considering their potential risks and unintended consequences.

Introduction

As the urgency of addressing climate change intensifies, geoengineering technologies, deliberate large-scale interventions in the Earth's climate system, are gaining increasing attention. While these technologies offer the potential to mitigate some of the effects of climate change, they also raise profound ethical questions about our relationship with the planet, the distribution of risks and benefits, and our responsibilities to future generations.

Types of Geoengineering Technologies

Geoengineering technologies fall into two main categories:

Solar Radiation Management (SRM): These techniques aim to reduce the amount of solar radiation absorbed by the Earth, thereby cooling the planet. Examples include injecting aerosols into the stratosphere to reflect sunlight or brightening marine clouds.

Carbon Dioxide Removal (CDR): These techniques aim to remove carbon dioxide from the atmosphere, either by enhancing natural processes (e.g., afforestation, ocean fertilization) or through technological means (e.g., direct air capture, bioenergy with carbon capture and storage).

Ethical Implications of Geoengineering

The ethical dilemmas associated with geoengineering are multifaceted and complex:

Moral Hazard: The prospect of geoengineering could reduce incentives to mitigate greenhouse gas emissions, leading to a reliance on technological fixes rather than addressing the root causes of climate change.
Unintended Consequences: Geoengineering technologies could have unforeseen and potentially harmful impacts on the climate system, ecosystems, and human societies.
Distributive Justice: The benefits and risks of geoengineering may not be distributed equitably, with some regions or populations benefiting more than others, or even experiencing negative impacts.
Intergenerational Justice: Deploying geoengineering technologies could impose risks and burdens on future generations, who will have to deal with the long-term consequences.
Hubris and Technological Control: Geoengineering raises questions about the extent to which humans should attempt to control the Earth's climate system, potentially leading to a sense of hubris and a disregard for the natural world.
Specific Ethical Considerations for SRM and CDR

SRM: Raises concerns about potential disruptions to regional weather patterns, impacts on the ozone layer, and the possibility of "termination shock" if deployment is suddenly stopped. It also raises questions about who controls the "global thermostat" and the potential for political conflict.
CDR: While generally considered less risky than SRM, CDR technologies also raise ethical questions about land use, resource consumption, and potential impacts on biodiversity.

Governance and Decision-Making

Given the potential global impacts of geoengineering, international cooperation and governance are essential. Key questions include:

Who decides whether and how to deploy geoengineering technologies?
How can we ensure transparency and public participation in decision-making?
How can we assess and manage the risks of geoengineering?
How can we ensure that geoengineering research and deployment are conducted ethically and responsibly?

Examples and Case Studies

The SPICE Project (UK): This research project, which explored the feasibility of injecting aerosols into the stratosphere, was canceled due to public concerns and ethical objections.
The Oxford Principles: These principles, developed by a group of experts, provide ethical guidelines for geoengineering research and governance.
The ongoing debate on ocean fertilization: This illustrates the ethical complexities of CDR technologies and the potential impacts on marine ecosystems.

Conclusion

Geoengineering technologies present a complex ethical landscape. While they offer a potential tool for addressing climate change, they also raise profound questions about our relationship with the planet, the distribution of risks and benefits, and our responsibilities to future generations. Careful consideration of these ethical dilemmas, along with robust governance and international cooperation, are essential to ensure that any decisions about geoengineering are made in the best interests of humanity and the planet.

Earth and Myth: When Geology Inspires Storytelling

Subtopic 1: Volcanic Fires and Creation Myths

Volcanic eruptions, with their awe-inspiring displays of fire, ash, and molten rock, have captivated human imagination for millennia. These geological events, both destructive and creative, have been deeply intertwined with the creation myths and stories of gods and goddesses in numerous cultures across the globe. From the fiery goddess Pele of Hawaiian mythology to the divine blacksmith Hephaestus of ancient Greece, volcanoes have inspired narratives that explain the origins of the world, the forces of nature, and the relationship between humans and the divine.

The Dance of Creation and Destruction: Pele, the Hawaiian Volcano Goddess

In the heart of the Pacific Ocean, the Hawaiian archipelago stands as a testament to the raw power of volcanic activity. For the indigenous people of Hawaii, volcanoes are not merely geological formations but the dwelling places of gods and the source of life itself. At the center of their cosmology is Pele, the goddess of fire, lightning, wind, and volcanoes.

Pele is revered as the creator of the Hawaiian islands, her fiery nature shaping the land with each eruption. She is often depicted as a passionate and unpredictable deity, her moods reflected in the volatile nature of volcanoes. When angered, she unleashes her fury in the form of lava flows and earthquakes, reminding mortals of the destructive power she wields. Yet, Pele is also a life-giver, her eruptions creating new land and enriching the soil, symbolizing the cyclical nature of creation and destruction.

The stories of Pele are deeply embedded in Hawaiian culture, serving as a reminder of the delicate balance between humans and the natural world. Her presence is felt in the vibrant landscapes of the islands, from the fiery glow of Kilauea's lava flows to the snow-capped summit of Mauna Kea. The reverence for Pele underscores the profound connection between the Hawaiian people and their volcanic environment, a relationship that has shaped their cultural identity for generations.

The Divine Smith: Hephaestus, the Greek God of Fire and Volcanoes

In the realm of ancient Greek mythology, Hephaestus, the god of fire, metalworking, and volcanoes, embodies the creative and destructive forces of the earth's interior. Born deformed and cast out of Mount Olympus by his mother, Hera, Hephaestus found refuge in the depths of the earth, where he established his forge.

Hephaestus's association with volcanoes is evident in his role as the divine blacksmith, forging weapons and armor for the gods in the fiery heart of volcanic mountains. His mastery over fire and metal symbolizes the transformative power of the earth's internal heat, shaping and molding the raw materials of the planet into objects of beauty and power.

The volcanic imagery associated with Hephaestus is further emphasized in his connection to the island of Lemnos, a volcanic island in the Aegean Sea. According to myth, Hephaestus landed on Lemnos after his fall from Olympus, and the island became his sacred abode. The volcanic landscape of Lemnos, with its fiery eruptions and rugged terrain, served as a fitting backdrop for the god of fire and volcanoes.

Volcanic Myths Across Cultures: A Tapestry of Fire and Creation

The influence of volcanic activity on mythology extends far beyond Hawaii and Greece, leaving its mark on cultures across the globe. In Iceland, the land of fire and ice, volcanoes are associated with the giants of Norse mythology, their fiery breath shaping the island's dramatic landscape. In Japan, Mount Fuji, a dormant volcano revered as a sacred mountain, is believed to be the abode of the goddess Konohanasakuya-hime, the deity of flowers and volcanoes.

In the Americas, the indigenous peoples of the Andes have long held a deep reverence for volcanoes, associating them with powerful mountain spirits. The Inca civilization, in particular, believed that volcanoes were the dwelling places of deities who controlled the forces of nature. Sacrifices and offerings were made to appease these mountain spirits and ensure the fertility of the land.

From the Pacific islands to the Andes mountains, volcanoes have inspired a rich tapestry of myths and legends, reflecting the profound impact of these geological events on human imagination. These stories, passed down through generations, continue to shape cultural beliefs and practices, reminding us of the awe-inspiring power of the earth's fiery heart.

Subtopic 2: Earthquakes and the Underworld

Earthquakes, with their sudden and violent shaking of the ground, have long been a source of fear and fascination for humanity. These seismic events, capable of toppling cities and reshaping landscapes in an instant, have inspired myths and legends about the underworld, subterranean creatures, and the fragility of the Earth's surface. Across cultures, earthquakes have been interpreted as manifestations of divine wrath, the restless movements of subterranean beings, or the Earth itself groaning under the weight of human transgressions.

The Wrath of the Gods: Earthquakes as Divine Punishment

In many ancient cultures, earthquakes were seen as a form of divine punishment, a way for the gods to express their displeasure with human behavior. In Greek mythology, Poseidon, the god of the sea and earthquakes, was known to shake the earth with his trident when angered, causing destruction and chaos. The ancient Greeks believed that earthquakes were a reminder of the gods' power and the need for humans to live in accordance with divine laws.

Similarly, in Norse mythology, earthquakes were attributed to the struggles of Loki, the trickster god, who was bound in chains beneath the earth. His attempts to break free were believed to cause tremors and earthquakes, signifying the potential for chaos and destruction that lay beneath the surface.

Subterranean Creatures and the Shaking Earth

In other cultures, earthquakes were attributed to the movements of creatures dwelling beneath the earth's surface. In Japanese mythology, Namazu, a giant catfish, was believed to cause earthquakes when it thrashed its tail. According to legend, Namazu was restrained by the god Kashima, but when Kashima's attention wavered, Namazu would move, causing the earth to tremble.

In Native American cultures, earthquakes were often associated with powerful spirits or creatures residing within the earth. The Cherokee people, for example, believed that the earth rested on the back of a giant turtle, and that earthquakes occurred when the turtle shifted its weight.

The Fragility of the Earth: Earthquakes as Reminders of Impermanence

Earthquakes also serve as a stark reminder of the fragility of the Earth's surface and the impermanence of human creations. In many cultures, myths and legends about earthquakes emphasize the destructive power of these events and the need for humans to respect the forces of nature.

In the Hindu epic Ramayana, an earthquake is described as shaking the earth and causing mountains to crumble, symbolizing the destructive power of nature and the impermanence of even the most seemingly solid structures. Similarly, in the Bible, earthquakes are often associated with divine judgment and the end of the world, highlighting the fragility of human existence in the face of natural forces.

Earthquakes in Modern Culture: From Disaster Narratives to Scientific Understanding

In modern times, earthquakes continue to inspire stories and narratives, reflecting our ongoing fascination with these powerful events. Disaster films and novels often depict the devastating consequences of earthquakes, exploring themes of survival, resilience, and the human response to catastrophe.

However, alongside these fictional accounts, scientific understanding of earthquakes has advanced significantly. Seismology, the study of earthquakes, has provided valuable insights into the causes and mechanisms of these events, allowing us to better predict and mitigate their impact.

Despite these advances, earthquakes remain a powerful reminder of the forces that shape our planet and the need for humans to live in harmony with the natural world. The myths and legends inspired by earthquakes continue to resonate with us, reminding us of the fragility of our existence and the awe-inspiring power of the earth beneath our feet.

Subtopic 3: Floods and the Deluge

Flood myths are a recurring motif in cultures across the globe, from the Biblical Flood to the Mesopotamian Epic of Gilgamesh. These narratives, often depicting a cataclysmic deluge that wipes out most of humanity, have captivated human imagination for millennia. While these stories are often imbued with religious and moral significance, they also raise intriguing questions about their potential connections to real geological events.

The Biblical Flood: A Story of Destruction and Renewal

The story of Noah's Ark and the Great Flood is one of the most well-known flood myths in the world. As recounted in the Book of Genesis, God, angered by the wickedness of humanity, decides to send a flood to cleanse the earth. Noah, a righteous man, is instructed to build an ark and take aboard his family and pairs of every animal species. The floodwaters rise, covering the entire earth and destroying all living things except those on the ark. After the flood recedes, Noah and his family emerge to repopulate the earth, marking a new beginning for humanity.

The Biblical Flood narrative has had a profound impact on Western culture and religious thought. It serves as a cautionary tale about the consequences of human sin and the importance of obedience to God. It also highlights themes of divine judgment, redemption, and the possibility of a new beginning after catastrophe.

The Epic of Gilgamesh: A Mesopotamian Flood Myth

The Epic of Gilgamesh, one of the oldest surviving works of literature, also features a prominent flood narrative. In this Mesopotamian epic, the hero Gilgamesh seeks immortality after the death of his friend Enkidu. He travels to meet Utnapishtim, a survivor of a great flood who has been granted eternal life. Utnapishtim recounts the story of the flood, which was sent by the gods to destroy humanity. He was instructed to build a boat and take aboard his family and possessions, as well as animals and craftsmen. After the flood subsides, Utnapishtim and his companions emerge to repopulate the earth.

The flood narrative in the Epic of Gilgamesh shares similarities with the Biblical Flood, suggesting a possible common origin or cross-cultural influence. Both stories depict a catastrophic flood sent by divine powers to punish humanity, and both feature a chosen individual who is instructed to build a vessel to survive the deluge.

Flood Myths Across Cultures: A Universal Theme

Flood myths are not limited to the Judeo-Christian and Mesopotamian traditions. Similar narratives can be found in cultures across the globe, from ancient Greece and India to China and the Americas. These stories often share common elements, such as a divine cause for the flood, a chosen survivor, and a vessel that carries humans and animals to safety.

The prevalence of flood myths across cultures suggests that they may reflect a shared human experience or a deep-seated fear of natural disasters. Some scholars have proposed that these myths may be based on real geological events, such as the flooding of the Black Sea basin around 5600 BCE, which could have displaced large populations and left a lasting impact on cultural memory.

Paleoclimatology and Flood Geology: Searching for Evidence

Paleoclimatology, the study of past climates, and flood geology, the study of ancient floods, have provided evidence of significant flooding events in various parts of the world. These events, caused by factors such as glacial melting, rising sea levels, and extreme rainfall, could have inspired the flood myths that have persisted through the ages.

For example, geological evidence suggests that the Black Sea basin was once a freshwater lake that was inundated by rising sea levels around 5600 BCE. This event, which could have caused widespread flooding and displacement of populations, has been proposed as a possible inspiration for the flood myths of the Near East.

The Enduring Power of Flood Myths

Flood myths continue to resonate with us today, not only as religious or cultural narratives but also as cautionary tales about the power of nature and the fragility of human civilization. In the face of climate change and the increasing frequency of extreme weather events, these stories serve as a reminder of the potential consequences of human actions and the need to live in harmony with the natural world.

Subtopic 4: The Enduring Power of Geo-mythology

Geo-mythology, the study of how geological events and phenomena have inspired myths, legends, and folklore, offers a fascinating window into the intersection of human culture and the natural world. These narratives, passed down through generations, reflect our ancestors' attempts to understand and explain the awe-inspiring forces that shape our planet. In the Anthropocene, an era defined by human impact on the environment, geo-mythology takes on renewed relevance, offering insights into our relationship with the Earth and inspiring new ways of thinking about environmental change and the future of our planet.

Shaping Cultural Beliefs, Values, and Practices

Geo-mythology has played a pivotal role in shaping cultural beliefs, values, and practices across the globe. Myths and legends about geological events often serve as:

Explanations of natural phenomena: In the absence of scientific understanding, myths provided explanations for earthquakes, volcanic eruptions, floods, and other geological events. These narratives often attributed these phenomena to the actions of gods, spirits, or other supernatural forces, shaping people's understanding of the world around them.

Moral and ethical frameworks: Many geo-myths convey moral and ethical lessons, highlighting the consequences of human actions and the importance of living in harmony with nature. For example, flood myths often serve as cautionary tales about the dangers of human hubris and the need for respect for the natural world.

Social cohesion and identity: Geo-mythology can contribute to social cohesion and identity by providing a shared narrative and cultural heritage. Myths and legends about geological events can create a sense of belonging and connection to a particular place and community.

Rituals and ceremonies: Many cultures have developed rituals and ceremonies associated with geological features and events. These practices often serve to appease deities, ensure good harvests, or protect communities from natural disasters.

Geo-mythology in the Anthropocene

In the Anthropocene, as we grapple with the consequences of human impact on the environment, geo-mythology takes on renewed significance. These ancient narratives can:

Foster environmental awareness: Geo-mythology can help us to reconnect with the natural world and appreciate the interconnectedness of all living things. By understanding how our ancestors perceived and interacted with the Earth, we can gain a deeper appreciation for the delicate balance of ecosystems and the importance of environmental stewardship.

Inspire climate action: Many geo-myths highlight the destructive potential of natural forces and the consequences of human actions that disrupt the balance of nature. These narratives can serve as powerful

reminders of the need for urgent action to address climate change and protect the planet for future generations.
Promote resilience and adaptation: Geo-mythology can offer insights into how past societies coped with environmental change and natural disasters. These stories can inspire resilience and adaptation in the face of the challenges posed by the Anthropocene.
Encourage interdisciplinary collaboration: Geo-mythology bridges the gap between the sciences and the humanities, encouraging interdisciplinary collaboration to address complex environmental issues. By bringing together diverse perspectives, we can develop more holistic and sustainable solutions.
Case Studies: Geo-mythology in Action

The Māori of New Zealand: The Māori people have a rich tradition of geo-mythology, with stories about the creation of the islands, volcanic eruptions, and earthquakes. These narratives are deeply intertwined with their cultural identity and their relationship with the land. In recent years, Māori knowledge and perspectives have been increasingly incorporated into environmental management and conservation efforts.
The Inuit of the Arctic: The Inuit people have a deep understanding of the Arctic environment, reflected in their myths and legends about animals, ice, and weather patterns. As the Arctic undergoes rapid change due to climate change, Inuit knowledge and geo-mythology are proving invaluable in adapting to new conditions and developing sustainable practices.
The Hawaiian Islands: The Hawaiian people have a strong connection to their volcanic landscape, reflected in their reverence for Pele, the volcano goddess. This cultural connection has influenced conservation efforts and disaster preparedness strategies, recognizing the power and significance of volcanic activity in shaping the islands.

Conclusion

Geo-mythology offers a powerful lens through which to explore the complex relationship between humans and the Earth. These ancient narratives, imbued with cultural meaning and ecological wisdom, can inspire us to reconnect with the natural world, address the challenges of the Anthropocene, and build a more sustainable future. By listening to the stories of the Earth, we can gain a deeper understanding of our place in the web of life and our responsibility to protect the planet for generations to come.

Rare Earths: The Hidden Pillars of Modern Civilization

Subtopic 1: Technological Enablers: Exploring the unique properties of rare earth elements and their critical role in modern technologies, including electronics, magnets, renewable energy, and medical devices.

Introduction

Rare earth elements (REEs) are a group of 17 elements that are essential for many modern technologies. They have unique properties that make them ideal for use in electronics, magnets, renewable energy, and medical devices.

Properties of Rare Earth Elements

REEs have a number of properties that make them valuable for use in technology. These include:

Luminescence: REEs can emit light when excited by an energy source. This property is used in fluorescent lights, lasers, and medical imaging.
Magnetism: REEs can be used to create powerful magnets. These magnets are used in wind turbines, electric vehicles, and medical devices.
Catalysis: REEs can be used to speed up chemical reactions. This property is used in catalytic converters, petroleum refining, and pollution control.
High Strength: REEs can be used to create strong, lightweight alloys. These alloys are used in aerospace, defense, and other high-performance applications.

Applications of Rare Earth Elements

REEs are used in a wide variety of technologies, including:

Electronics: REEs are used in smartphones, computers, and other electronic devices. They are also used in the production of semiconductors and flat-panel displays.
Magnets: REEs are used to create powerful magnets that are used in wind turbines, electric vehicles, and medical devices.
Renewable Energy: REEs are used in solar panels, wind turbines, and other renewable energy technologies.
Medical Devices: REEs are used in medical imaging, cancer treatment, and other medical applications.
Examples of Rare Earth Elements in Technology

Europium: Europium is used in the red phosphor in color televisions and computer monitors.
Neodymium: Neodymium is used to create powerful magnets that are used in wind turbines and electric vehicles.
Yttrium: Yttrium is used in high-temperature superconductors and lasers.
Cerium: Cerium is used in catalytic converters and polishing compounds.
Case Studies

The Role of Rare Earth Elements in the Development of Electric Vehicles: REEs are essential for the development of electric vehicles. They are used in the batteries, motors, and generators of electric vehicles.
The Use of Rare Earth Elements in Medical Imaging: REEs are used in medical imaging to create contrast agents that help doctors to see inside the body.
The Impact of Rare Earth Elements on the Renewable Energy Industry: REEs are essential for the development of renewable energy technologies. They are used in solar

panels, wind turbines, and other renewable energy technologies.

Conclusion

REEs are essential for many modern technologies. They have unique properties that make them ideal for use in electronics, magnets, renewable energy, and medical devices. The demand for REEs is expected to continue to grow in the coming years.

Subtopic 2: Uneven Distribution and Geopolitics: Analyzing the global distribution of rare earth deposits, the challenges of extraction and processing, and the geopolitical implications of resource scarcity and dependence.

The story of rare earth elements is inextricably linked to the Earth's geological history, a narrative spanning billions of years. These elements, despite their name, are relatively abundant in the Earth's crust. However, they are rarely found in concentrated, economically exploitable deposits. This uneven distribution has shaped the geopolitical landscape, creating dependencies and vulnerabilities that nations and industries are grappling with today.

Global Distribution: A Geological Lottery

The concentration of rare earths is a result of complex geological processes that have unfolded over vast stretches of time. While rare earths are found in trace amounts throughout the Earth's crust, economically viable deposits are concentrated in specific regions.

China: Holds the lion's share of global rare earth reserves and has dominated production for decades. The Bayan Obo mining district in Inner Mongolia is a prime example,

hosting one of the world's largest and richest rare earth deposits.
Vietnam: Emerging as a significant player, with substantial reserves located in the northwestern region.
Brazil: Home to significant deposits, particularly in the Amazon region, though exploitation faces environmental and logistical challenges.
Australia: Possesses substantial reserves, with Mount Weld in Western Australia being a key source of heavy rare earths.
United States: While holding considerable reserves, particularly in California and Alaska, domestic production has been limited in recent years.
This uneven distribution has led to a situation where a few countries hold a dominant position in the rare earth supply chain. This concentration of power has raised concerns about supply security and the potential for resource nationalism.

Challenges of Extraction and Processing

Extracting and processing rare earths is a complex and often environmentally damaging endeavor. The challenges include:

Complex Mineralogy: Rare earth ores often contain a mixture of different rare earth elements, making separation and purification a complex and energy-intensive process.
Radioactive Contamination: Some rare earth ores contain radioactive elements like thorium and uranium, posing environmental and health risks.
Environmental Impacts: Mining and processing can lead to deforestation, habitat destruction, water pollution, and the release of toxic chemicals.

These challenges contribute to the high cost of rare earth production and raise concerns about the sustainability of the industry.

Geopolitical Implications: A New Great Game?

The strategic importance of rare earths has led to a renewed focus on resource security and the potential for geopolitical competition.

China's Dominance: China's dominance in rare earth production has given it significant leverage in the global market. This has led to concerns about supply disruptions and price volatility.
Diversification Efforts: Countries and companies are actively seeking to diversify their rare earth supply chains by investing in new mining projects, developing alternative materials, and promoting recycling.
Trade Tensions: The rare earth industry has become entangled in trade disputes, with countries imposing tariffs and quotas on rare earth exports.
Technological Competition: Rare earths are critical for many emerging technologies, including renewable energy and electric vehicles. This has led to concerns about a "rare earth race" and the potential for resource conflicts.

Case Studies

The 2010 China-Japan Rare Earth Dispute: This dispute highlighted the vulnerability of global supply chains to geopolitical tensions. China's restriction of rare earth exports to Japan sent shockwaves through the global market and spurred diversification efforts.
The Mountain Pass Mine in the United States: The reopening of the Mountain Pass mine in California is an example of efforts to revive domestic rare earth production and reduce reliance on imports.

Lynas Corporation in Malaysia: The Lynas Advanced Materials Plant in Malaysia has faced opposition due to concerns about radioactive waste disposal, illustrating the environmental challenges associated with rare earth processing.

Conclusion

The uneven distribution of rare earths has created a complex geopolitical landscape, with countries vying for access to these critical resources. The challenges of extraction and processing, coupled with the growing demand for rare earths, underscore the need for sustainable and responsible resource management. The future of the rare earth industry will depend on finding a balance between economic development, environmental protection, and geopolitical stability.

Subtopic 3: The Environmental Cost: Investigating the environmental impacts of rare earth mining and processing, including habitat destruction, water pollution, and the release of radioactive materials.

The quest for rare earths, these technological cornerstones of modern civilization, comes at a cost. The extraction and processing of these elements leave a significant environmental footprint, impacting ecosystems, communities, and the planet as a whole. Understanding these impacts is crucial for developing sustainable practices and mitigating the environmental risks associated with rare earth production.

Environmental Impacts of Rare Earth Mining

Rare earth mining, like any large-scale mining operation, can have significant environmental consequences:

Habitat Destruction: Open-pit mining, the most common method used for rare earth extraction, involves removing vast amounts of topsoil and vegetation, leading to deforestation, habitat loss, and disruption of ecosystems.
Land Degradation: Mining activities can lead to soil erosion, land subsidence, and the formation of sinkholes, altering the landscape and impacting land use.
Dust Pollution: Mining operations generate significant amounts of dust, which can contain heavy metals and radioactive particles. This dust can contaminate air and water, posing risks to human health and the environment.
Environmental Impacts of Rare Earth Processing

The processing of rare earth ores is even more environmentally damaging than the mining itself. The challenges include:

Water Pollution: Processing rare earths requires large amounts of water and generates significant volumes of wastewater, which can be contaminated with acids, heavy metals, and radioactive materials. This wastewater can pollute rivers, lakes, and groundwater, harming aquatic life and threatening human health.
Radioactive Waste: Many rare earth ores contain radioactive elements like thorium and uranium. The processing of these ores generates radioactive waste, which needs to be carefully managed to prevent environmental contamination and health risks.
Air Pollution: Processing facilities release air pollutants, including sulfur dioxide, nitrogen oxides, and particulate matter, contributing to smog and acid rain.
Chemical Contamination: The use of strong acids and other chemicals in the separation and purification of rare earths can lead to soil and water contamination, impacting ecosystems and human health.

Case Studies

The Bayan Obo Mining District in China: This massive mining operation has been linked to significant environmental problems, including water pollution, air pollution, and radioactive contamination. The surrounding environment has been severely degraded, with impacts on human health and livelihoods.
The Lynas Advanced Materials Plant in Malaysia: This rare earth processing facility has faced controversy due to concerns about radioactive waste disposal and potential environmental contamination. The case highlights the challenges of managing the environmental risks associated with rare earth processing.
The Impact of Rare Earth Mining on Biodiversity: A study in Vietnam found that rare earth mining has led to the loss of forest habitat and the decline of several endangered species, including primates and birds. This highlights the need to protect biodiversity in regions where rare earth mining takes place.

Mitigating the Environmental Impacts

Addressing the environmental challenges of rare earth production requires a multi-faceted approach:

Sustainable Mining Practices: Implementing sustainable mining practices, such as reducing water and energy consumption, minimizing waste generation, and1 restoring mined areas, can help to reduce the environmental footprint of rare earth mining.

Improved Processing Technologies: Developing cleaner and more efficient processing technologies can reduce water and energy consumption, minimize waste generation, and reduce the release of pollutants.
Recycling and Reuse: Recycling and reusing rare earths can reduce the demand for primary production and

minimize the environmental impacts associated with mining and processing.
Environmental Regulations: Strong environmental regulations and enforcement are essential to ensure that rare earth mining and processing operations meet environmental standards and minimize their impacts on ecosystems and communities.

Conclusion

The environmental costs of rare earth production are significant and cannot be ignored. The impacts of mining and processing on ecosystems, communities, and human health underscore the urgent need for sustainable practices and responsible resource management. By adopting cleaner technologies, promoting recycling, and strengthening environmental regulations, we can strive to minimize the environmental footprint of rare earths and ensure a more sustainable future for this critical industry.

Subtopic 4: Towards a Circular Economy:

Exploring strategies for reducing our reliance on rare earths, including recycling, developing alternative materials, and promoting sustainable consumption patterns.

The linear "take-make-dispose" model that has characterized industrial production for decades is increasingly unsustainable, especially when it comes to critical resources like rare earths. The environmental and geopolitical challenges associated with rare earth extraction and processing highlight the urgent need for a transition to a circular economy. This approach aims to keep resources in use for as long as possible, extracting maximum value from them while minimizing waste and environmental impact.

Recycling: Closing the Loop

Recycling plays a crucial role in the circular economy for rare earths. It allows us to recover valuable materials from end-of-life products, reducing the demand for primary production and minimizing the environmental burden of mining and processing.

Challenges: Recycling rare earths is technically challenging and often economically unviable due to the low concentrations of these elements in many products. The complex separation and purification processes required can be energy-intensive and costly.
Opportunities: Advances in recycling technologies, such as hydrometallurgy and bioleaching, are making it possible to recover rare earths from a wider range of products, including electronic waste, batteries, and magnets.
Examples:
Urban Mining: Extracting valuable materials, including rare earths, from electronic waste and other urban waste streams.
Closed-Loop Recycling: Designing products for disassembly and recycling, allowing for the recovery of rare earths and other components for reuse in new products.
Extended Producer Responsibility: Holding manufacturers responsible for the end-of-life management of their products, incentivizing them to design for recyclability and invest in recycling infrastructure.
Developing Alternative Materials

Another key strategy for reducing our reliance on rare earths is to develop alternative materials that can perform similar functions. This can involve:

Material Substitution: Replacing rare earth elements with more abundant and readily available materials in certain applications.
Material Efficiency: Designing products to use less material overall, reducing the demand for rare earths and other resources.
Innovation in Material Science: Developing new materials with enhanced properties that can outperform rare earths in specific applications.
Examples:
Electric Vehicle Motors: Researchers are exploring the use of alternative materials, such as ferrite magnets and copper-iron alloys, to reduce or eliminate the need for rare earths in electric vehicle motors.
Wind Turbine Generators: Developing direct-drive wind turbines that use less rare earth magnets or exploring alternative generator designs that rely on different technologies.
Lighting Technologies: LED lighting, which is more energy-efficient and uses less rare earths than traditional fluorescent lighting, is becoming increasingly prevalent.
Promoting Sustainable Consumption Patterns

Shifting towards a circular economy also requires changes in consumption patterns. This involves:

Product Life Extension: Encouraging the repair, reuse, and refurbishment of products to extend their lifespan and reduce the need for new production.
Sharing Economy: Promoting sharing platforms and services that allow people to access goods and services without owning them, reducing the demand for material resources.
Consumer Awareness: Educating consumers about the environmental and social impacts of their consumption choices and encouraging them to make more sustainable decisions.

Examples:
Right to Repair: Advocating for policies that make it easier for consumers to repair their own products, extending their lifespan and reducing electronic waste.
Product-as-a-Service: Shifting from selling products to providing services, where companies retain ownership of the product and are responsible for its maintenance and end-of-life management.
Circular Design: Designing products with a focus on durability, repairability, and recyclability, promoting longer product lifespans and facilitating material recovery.
Case Studies

The European Union's Circular Economy Action Plan: This ambitious plan sets out a comprehensive framework for transitioning to a circular economy, including targets for recycling, waste reduction, and sustainable product design.
Japan's "Sound Material-Cycle Society": Japan has been a pioneer in promoting a circular economy, with policies and initiatives aimed at reducing waste, promoting recycling, and developing innovative technologies for resource recovery.
The Ellen MacArthur Foundation: This non-profit organization is a leading advocate for the circular economy, working with businesses, governments, and academia to promote circular economy principles and practices.
Conclusion

Transitioning to a circular economy for rare earths is essential for ensuring a sustainable and secure future for these critical resources. By embracing recycling, developing alternative materials, and promoting sustainable consumption patterns, we can reduce our reliance on primary production, minimize environmental impacts, and create a more resilient and equitable economy. The journey towards a circular economy will require collaboration between governments, businesses, and consumers, but the rewards are significant: a healthier planet, a more secure resource base, and a more prosperous future for all.

Gazing into the Abyss: Predicting the Deep Future of Earth

Subtopic 1: The Dance of Continents: Projecting the future movements of Earth's tectonic plates and the long-term reshaping of continents, oceans, and mountain ranges.

The Ever-Shifting Jigsaw

Imagine Earth as a giant jigsaw puzzle, its pieces constantly in motion, colliding, separating, and grinding against each other. These pieces are the tectonic plates, massive slabs of Earth's lithosphere that float on the semi-fluid asthenosphere beneath. Their slow but relentless dance, driven by convection currents deep within the Earth, has shaped our planet's surface for billions of years, creating continents, oceans, and mountain ranges.

A Glimpse into the Future

Predicting the future movements of these plates is a complex task, but scientists use various tools and techniques to peer into the deep future of Earth's geography. One approach is geodynamic modeling, which uses computer simulations to project plate movements based on our understanding of plate tectonics and the forces that drive them. These models take into account factors such as plate velocities, the strength of plate boundaries, and the distribution of heat within the Earth.

Another approach is to study the past. By examining the geological record, scientists can reconstruct the history of plate movements and identify patterns that may help predict future trends. This field, known as paleogeography, provides valuable insights into the long-term evolution of Earth's surface.

A World Transformed

Based on these approaches, scientists have painted a picture of Earth's future geography, a world transformed by the ongoing dance of continents. Here are some of the key predictions:

The Closing of the Atlantic: The Atlantic Ocean, which has been steadily widening for millions of years, is projected to begin closing in the distant future. The Americas will eventually collide with Africa and Eurasia, forming a new supercontinent, often referred to as "Pangaea Proxima" or "Amasia."

The Rise of a New Mountain Range: The collision of the Americas with Africa and Eurasia will also lead to the formation of a massive mountain range, rivaling the Himalayas in scale. This new mountain range will stretch along the eastern coast of North and South America, creating a dramatic new landscape.

The Shrinking of the Pacific: As the Atlantic closes, the Pacific Ocean will continue to shrink, eventually becoming a remnant of its former self. The Pacific Ring of Fire, a zone of intense volcanic and seismic activity, will also gradually diminish.

The Formation of New Oceans: As continents collide and separate, new oceans will form in their wake. These new oceans will create new habitats for marine life and influence global climate patterns.

Examples and Case Studies

The Mediterranean Sea: The Mediterranean Sea is a remnant of the Tethys Ocean, which once separated Africa and Eurasia. The ongoing collision of these continents is slowly closing the Mediterranean, and it is projected to eventually disappear, leaving behind a mountain range in its place.

The Himalayas: The Himalayas, the highest mountain range on Earth, were formed by the collision of India with Asia, which began about 50 million years ago. This collision is still ongoing, and the Himalayas continue to rise at a rate of about 5 millimeters per year.

The East African Rift: The East African Rift is a zone of active continental rifting, where the African plate is splitting apart. This rifting is creating a new ocean basin, which will eventually separate East Africa from the rest of the continent.

The Human Impact

The future movements of Earth's tectonic plates will have profound implications for human civilization. The formation of new continents, oceans, and mountain ranges will reshape coastlines, alter climate patterns, and create new challenges and opportunities for human societies. Understanding these long-term changes is crucial for planning for the future and ensuring the sustainability of our planet.

Subtopic 2: A Changing Climate (Continued): Modeling future climate scenarios based on different greenhouse gas emission pathways, exploring the potential impacts on sea level rise, extreme weather events, and the habitability of different regions.

The Unfolding Climate Story

Earth's climate is a complex and dynamic system, influenced by a multitude of factors, including solar radiation, atmospheric composition, and ocean currents. While it has always naturally fluctuated, human activities, particularly the burning of fossil fuels, have dramatically altered the delicate balance, leading to unprecedented warming. Understanding how this warming will unfold in the future is crucial for mitigating its impacts and adapting to a changing world.

Climate Modeling: Peering into the Future

Scientists use sophisticated computer models, known as Global Climate Models (GCMs), to project future climate scenarios. These models simulate the Earth's climate system, taking into account various factors such as atmospheric circulation, ocean currents, and land surface processes. By inputting different greenhouse gas emission pathways, representing different societal choices and policies, these models can generate a range of possible futures.

The Intergovernmental Panel on Climate Change (IPCC), the leading international body for assessing climate change, uses these models to develop a set of standardized scenarios, known as Representative Concentration Pathways (RCPs). These scenarios range from RCP2.6, representing a pathway with strong mitigation efforts and low greenhouse gas emissions, to RCP8.5, representing a business-as-usual scenario with high emissions.

The Impacts: A Cascade of Changes

The projected impacts of climate change are far-reaching and multifaceted, affecting every corner of the globe. Some of the key impacts include:

Sea Level Rise: As the planet warms, glaciers and ice sheets melt, and ocean water expands, leading to rising sea levels. This poses a significant threat to coastal

communities, increasing the risk of flooding, erosion, and saltwater intrusion into freshwater sources.
Extreme Weather Events: Climate change is expected to increase the frequency and intensity of extreme weather events, such as heat waves, droughts, floods, and storms. These events can have devastating impacts on human lives, infrastructure, and ecosystems.
Changes in Habitability: Rising temperatures, changing precipitation patterns, and extreme weather events will affect the habitability of different regions. Some areas may become too hot or dry to support human life, while others may experience more frequent and severe natural disasters.
Examples and Case Studies

The Maldives: This low-lying island nation in the Indian Ocean is particularly vulnerable to sea level rise. Even a small increase in sea level could inundate much of the country, displacing its population and threatening its existence.
California: This state in the western United States is experiencing an increase in wildfires due to hotter and drier conditions. These wildfires have caused widespread destruction, loss of life, and air pollution.
The Arctic: The Arctic is warming at a rate twice as fast as the global average, leading to the rapid melting of sea ice and permafrost. This is affecting Arctic ecosystems, threatening wildlife, and releasing methane, a potent greenhouse gas, into the atmosphere.
The Human Element

The future of Earth's climate is not predetermined. Human choices and actions will play a crucial role in shaping the trajectory of climate change. By reducing greenhouse gas emissions, investing in adaptation measures, and transitioning to a sustainable energy system, we can mitigate the impacts of climate change and create a more resilient future for all.

Subtopic 3: The Sun's Long Farewell: Examining the Sun's inevitable evolution into a red giant star and its implications for Earth's long-term habitability, including rising temperatures, ocean evaporation, and the potential engulfment of the planet.

Our Star's Inevitable Transformation

The Sun, our life-giving star, is not immutable. Like all stars, it is undergoing a slow but inexorable evolution, fueled by the nuclear fusion reactions at its core. While it currently provides a stable source of energy for Earth, its future holds dramatic changes that will ultimately render our planet uninhabitable.

The Red Giant Phase

In about 5 billion years, the Sun will exhaust the hydrogen fuel at its core and begin to burn helium. This will cause the Sun to expand dramatically, transforming it into a red giant star. Its outer layers will swell outwards, engulfing Mercury and Venus, and possibly even Earth.

As the Sun expands, its luminosity will increase, bathing Earth in intense heat. The oceans will begin to evaporate, creating a thick, steamy atmosphere. The surface temperature will soar, making the planet uninhabitable for most forms of life.

Earth's Uncertain Fate

Whether Earth will be completely engulfed by the Sun or narrowly escape its fiery embrace is still a matter of debate among scientists. Some models suggest that

Earth's orbit may expand as the Sun loses mass, potentially allowing it to avoid engulfment. However, even if Earth survives this initial phase, it will be a scorched, lifeless world, its surface a barren wasteland.

The Aftermath

After the red giant phase, the Sun will shed its outer layers, leaving behind a dense, hot core known as a white dwarf. This remnant star will gradually cool and fade over trillions of years, leaving Earth as a frozen, dark world orbiting a dying ember.

Implications for Life

The Sun's evolution into a red giant has profound implications for the future of life on Earth. While humans may not be around to witness this event, it is a reminder of the impermanence of our planet and the need to consider the long-term sustainability of our species.

Examples and Case Studies

Other Star Systems: Astronomers have observed red giant stars in other star systems, providing insights into the Sun's future evolution. These observations have revealed the dramatic changes that occur as stars age and the potential impacts on their planetary systems.
Planetary Nebulae: Planetary nebulae are the glowing remnants of stars that have shed their outer layers. These nebulae offer a glimpse into the Sun's distant future, when it will expel its outer layers, leaving behind a white dwarf.

The Search for New Homes

The Sun's eventual demise highlights the importance of exploring and potentially colonizing other planets. While this may seem like science fiction, it is a long-term goal that could ensure the survival of our species beyond the lifespan of our own star.

Subtopic 4: Earth's Ultimate Fate: Exploring the possible scenarios for Earth's ultimate fate, from a fiery demise within the Sun to a frozen, lifeless world orbiting a white dwarf star, considering the timescales involved and the potential for human intervention.

The Inevitable End

While the Earth has been a cradle for life for billions of years, its existence, like all things in the universe, is finite. The ultimate fate of our planet is intertwined with the life cycle of our Sun, and while the timeline stretches far beyond human comprehension, the end is inevitable. Let's explore the possible scenarios and the vast timescales involved.

Scenario 1: Engulfment by the Sun

As discussed in the previous subtopic, the Sun's transformation into a red giant poses the most immediate threat to Earth. In approximately 5 billion years, the Sun's expanding outer layers could engulf Earth, vaporizing its atmosphere and melting its surface. This would be a swift and dramatic end to our planet.

Scenario 2: A Scorched, Lifeless World

Even if Earth narrowly escapes being engulfed by the red giant Sun, the intense heat and radiation will render it uninhabitable. The oceans will boil away, the atmosphere will be stripped away, and the surface will become a molten wasteland. Earth will remain a planet, but a lifeless one, orbiting a dying star.

Scenario 3: A Frozen, Dark World

After the red giant phase, the Sun will shed its outer layers, leaving behind a white dwarf. This remnant star will gradually cool and fade, emitting little light or heat. Earth, if it survives the red giant phase, will become a frozen, dark world, its surface locked in a perpetual ice age.

Scenario 4: Ejected from the Solar System

While less likely, there's a possibility that gravitational interactions with other stars or massive objects passing through the solar system could perturb Earth's orbit, potentially ejecting it from the solar system altogether. Earth would become a rogue planet, wandering the vast emptiness of interstellar space, a cold, dark, and lonely world.

The Timescales Involved

These scenarios unfold over immense timescales that dwarf human history. The red giant phase is billions of years away, and the white dwarf phase will last for trillions of years. These timescales are so vast that they are difficult to grasp, but they underscore the transient nature of our existence in the cosmos.

The Potential for Human Intervention

While the ultimate fate of Earth may seem predetermined, there is a glimmer of hope in the potential for human intervention. Advanced civilizations, if they survive that long, could develop technologies to alter Earth's orbit, shield it from the Sun's radiation, or even relocate the planet to a safer region of the galaxy. These possibilities may seem like science fiction today, but they highlight the potential for human ingenuity to overcome seemingly insurmountable challenges.

Examples and Case Studies

Kepler-1658b: This exoplanet provides a glimpse into the potential fate of planets orbiting stars nearing the end of their lives. Kepler-1658b is spiraling towards its host star, a subgiant, and is expected to be engulfed within 3 million years.
White Dwarf Stars with Planetary Debris: Astronomers have observed white dwarf stars with remnants of planetary systems, providing evidence of the long-term fate of planets orbiting stars that have exhausted their nuclear fuel.

The Importance of Perspective

Contemplating the ultimate fate of Earth can be a humbling experience. It reminds us of our place in the vast universe and the fragility of our existence. However, it can also inspire us to cherish our planet, to strive for sustainability, and to explore the cosmos in search of new homes for life.

Conclusion

This journey through the deep future of Earth has taken us from the shifting continents to the dying embers of our Sun. We have explored the forces that shape our planet, the challenges that lie ahead, and the potential for human intervention. While the ultimate fate of Earth may be inevitable, the story is far from over. The future holds countless possibilities, and the choices we make today will determine the course of our planet and the destiny of our species.

Alternative Earths: Reimagining the Trajectory of Civilizations

Subtopic 1: The Butterfly Effect: Exploring the concept of historical contingency and how seemingly small changes in Earth's history or human choices could have led to dramatically different outcomes for civilizations.

(Research: Chaos theory, historical counterfactuals, alternative history)

Introduction

Imagine a world where the Roman Empire never fell, where the Americas were never colonized by Europeans, or where the Industrial Revolution never happened. These are just a few examples of the countless alternative histories that could have unfolded if even the slightest of changes had occurred in the past. The concept of the "Butterfly Effect," derived from chaos theory, suggests that even seemingly insignificant events can have a cascading impact on the course of history, leading to dramatically different outcomes. In this section, we will delve into the fascinating world of historical contingency, exploring how small changes in Earth's history or human choices could have led to entirely different trajectories for civilizations.

Chaos Theory and Historical Contingency

Chaos theory is a branch of mathematics that studies complex systems that are highly sensitive to initial conditions. This sensitivity means that even tiny changes in the starting conditions of a system can lead to vastly different outcomes over time. This concept is often illustrated by the metaphor of a butterfly flapping its wings in Brazil causing a tornado in Texas.

When applied to history, chaos theory suggests that even seemingly insignificant events or choices can have a profound impact on the course of civilizations. This idea is known as historical contingency, which emphasizes the role of chance and randomness in shaping the past. In essence, history is not a predetermined path but rather a series of contingent events that could have unfolded in countless different ways.

Historical Counterfactuals

One way to explore the concept of historical contingency is through the use of historical counterfactuals, also known as "what if?" scenarios. These are thought experiments that imagine how history might have unfolded if certain events had happened differently or not at all. By exploring these alternative histories, we can gain a deeper appreciation for the fragility of civilizations and the role of chance in shaping our world.

Some examples of historical counterfactuals include:

What if the Spanish Armada had succeeded in invading England in 1588?
What if the American Revolution had failed?
What if World War I had never happened?
These are just a few of the countless "what if?" questions that historians and writers have explored over the years. By considering these alternative scenarios, we can gain a better understanding of the factors that led to the world we live in today.

Alternative History

Alternative history is a genre of fiction that explores how history might have unfolded differently if certain events had occurred differently or not at all. This genre often uses historical counterfactuals as a starting point, but it also allows for more creative speculation about the past.

Some popular examples of alternative history include:

The Man in the High Castle by Philip K. Dick, which imagines a world where the Axis powers won World War II.
11/22/63 by Stephen King, which explores the possibility of preventing the assassination of John F. Kennedy.
The Years of Rice and Salt by Kim Stanley Robinson, which imagines a world where the Black Death wiped out 99% of Europe's population.
Alternative history can be a powerful tool for exploring the concept of historical contingency and the fragility of civilizations. It can also help us to appreciate the importance of human choices and the potential for shaping a better future.

Examples and Case Studies

The Battle of Salamis (480 BC): This naval battle between the Greeks and the Persians was a pivotal moment in ancient history. If the Persians had won, they might have conquered all of Greece, potentially altering the course of Western civilization.
The Mongol Invasions (13th century): The Mongol conquests had a devastating impact on much of Asia and Europe. If the Mongols had continued their westward expansion, they might have conquered all of Europe, leading to a very different world today.
The Black Death (14th century): This pandemic killed an estimated 30-60% of Europe's population, leading to significant social, economic, and cultural changes. If the Black Death had not occurred, Europe might have developed very differently.

The Industrial Revolution (18th-19th centuries): This period of technological innovation transformed the world, leading to unprecedented economic growth and social change. If the Industrial Revolution had not occurred, the world today would be a very different place.
Conclusion

The concept of the Butterfly Effect and historical contingency reminds us that the past is not a fixed and immutable entity. Even the smallest of changes could have led to vastly different outcomes for civilizations. By exploring historical counterfactuals and alternative histories, we can gain a deeper appreciation for the fragility of our world and the importance of human choices in shaping the future.

Subtopic 2: Geological Turning Points: Analyzing how major geological events, such as volcanic super-eruptions, asteroid impacts, or shifts in ocean currents, could have altered the course of human history and the development of civilizations.

(Research: Volcanology, impact events, paleoclimatology, historical analysis)

Introduction

While human choices and actions undoubtedly play a significant role in shaping history, it is crucial to acknowledge the profound impact of geological events on the trajectory of civilizations. From volcanic super-eruptions that darken the skies and trigger volcanic winters to asteroid impacts that cause mass extinctions, these powerful forces of nature have the potential to reshape the planet and alter the course of human history in profound ways. In this section, we will explore how major geological events could have altered the course of human history and the development of civilizations.

Volcanic Super-Eruptions

Volcanic super-eruptions are among the most catastrophic natural events on Earth. These eruptions can eject massive amounts of ash and gas into the atmosphere, blocking sunlight and causing global cooling, a phenomenon known as a volcanic winter. The effects of a volcanic winter can be devastating, leading to crop failures, famine, and widespread social unrest.

One of the most well-known super-eruptions is the Toba eruption in Indonesia, which occurred approximately 74,000 years ago. This eruption is believed to have caused a global volcanic winter that lasted for several years, leading to a significant decline in human populations. Some scientists even suggest that the Toba eruption may have brought humanity to the brink of extinction.

If a super-eruption were to occur today, the consequences would be catastrophic. Modern civilization is highly dependent on global trade and transportation networks, which would be severely disrupted by a volcanic winter. The resulting food shortages and economic collapse could lead to widespread conflict and social upheaval.

Asteroid Impacts

Asteroid impacts are another type of geological event that can have a profound impact on human history. The Chicxulub impact, which occurred approximately 66 million years ago, is believed to have caused the extinction of the dinosaurs 1 and paved the way for the rise of mammals.

While smaller asteroid impacts are relatively common, a large impact could have devastating consequences for human civilization. A large asteroid impact could trigger tsunamis, earthquakes, and wildfires, causing widespread destruction and loss of life. The impact could also eject large amounts of dust and debris into the atmosphere, blocking sunlight and causing a global impact winter.

The risk of a large asteroid impact is relatively low, but the potential consequences are so severe that it is important to take this threat seriously. Scientists are constantly monitoring the skies for potentially hazardous asteroids, and there are ongoing efforts to develop technologies that could deflect an asteroid on a collision course with Earth.

Shifts in Ocean Currents

Ocean currents play a crucial role in regulating Earth's climate. Changes in ocean currents can lead to significant shifts in temperature and precipitation patterns, which can have a profound impact on human societies.

One example of a major shift in ocean currents is the Younger Dryas event, which occurred approximately 12,900 years ago. This event is believed to have been caused by a sudden influx of freshwater into the North Atlantic Ocean, which disrupted the thermohaline circulation. The Younger Dryas event led to a sudden cooling of the Northern Hemisphere, which may have contributed to the extinction of many large mammals and the decline of some human cultures.

Changes in ocean currents can also have a significant impact on human societies today. For example, the El Niño-Southern Oscillation (ENSO) is a climate pattern that can cause droughts in some regions and floods in others. ENSO events can have a significant impact on agriculture, water resources, and human health.

Examples and Case Studies

The Minoan Eruption (1628 BC): This volcanic eruption on the island of Thera (modern Santorini) is believed to have caused a tsunami that devastated the Minoan civilization on the island of Crete. The eruption may also have contributed to the decline of the Minoan civilization.
The Tambora Eruption (1815): This eruption in Indonesia was the largest volcanic eruption in recorded history. The eruption caused a global volcanic winter, leading to the "Year Without a Summer" in 1816. The resulting crop failures and famine led to widespread social unrest.
The Little Ice Age (1300-1850): This period of cooling is believed to have been caused by a combination of factors, including volcanic eruptions and changes in solar activity. The Little Ice Age had a significant impact on human societies, leading to crop failures, famine, and disease outbreaks.

Conclusion

Geological events have played a significant role in shaping the course of human history and the development of civilizations. From volcanic super-eruptions to asteroid impacts to shifts in ocean currents, these powerful forces of nature have the potential to reshape the planet and alter the trajectory of human societies. While we cannot prevent these events from happening, we can study them and learn from them, so that we can be better prepared for the future.

Subtopic 3: The Road Not Taken: Imagining alternative scenarios for the rise and fall of civilizations based on different environmental conditions, technological innovations, or social structures.

(Research: World-building, speculative fiction, historical analysis, social modeling)

Introduction

History is a tapestry woven from countless threads of cause and effect, where environmental conditions, technological advancements, and social structures intertwine to shape the destinies of civilizations. But what if we could unravel these threads and reweave them into alternative patterns, exploring the "roads not taken" in the grand narrative of human history? This section invites us to embark on a journey of imagination, envisioning alternative scenarios for the rise and fall of civilizations based on different environmental conditions, technological innovations, or social structures.

Environmental Divergence

Imagine a world where the Sahara Desert remained a lush grassland, teeming with life and supporting thriving civilizations. Or perhaps a world where the Amazon rainforest never existed, replaced by vast savannas that fostered the development of nomadic horse cultures. These are just a few examples of how different environmental conditions could have dramatically altered the course of human history.

Environmental factors, such as climate, geography, and natural resources, play a crucial role in shaping the development of civilizations. The availability of fertile land, water resources, and raw materials can determine the size and complexity of societies, while climate change can trigger migrations, conflicts, and even the collapse of entire civilizations.

By exploring alternative environmental scenarios, we can gain a deeper appreciation for the delicate balance between human societies and the natural world. We can also consider how different environmental conditions might have led to the emergence of unique cultures, technologies, and social structures.

Technological Turning Points

Imagine a world where the steam engine was never invented, where electricity remained a scientific curiosity, or where the internet never came to be. These are just a few examples of how different technological innovations could have profoundly altered the course of human history.

Technological advancements have been a driving force behind the rise and fall of civilizations throughout history. From the invention of agriculture to the development of the printing press, new technologies have transformed human societies, enabling them to grow, prosper, and expand their influence.

By exploring alternative technological scenarios, we can consider how different inventions or discoveries might have led to different social, economic, and political structures. We can also contemplate the potential benefits and risks of technological progress, and how it can be harnessed to create a more sustainable and equitable future.

Social Structures and Cultural Values

Imagine a world where women held positions of power throughout history, where slavery was never institutionalized, or where democracy emerged in ancient Mesopotamia instead of ancient Greece. These are just a few examples of how different social structures and cultural values could have shaped the course of human history.

Social structures and cultural values play a crucial role in shaping the development of civilizations. The way societies organize themselves, distribute resources, and make decisions can have a profound impact on their stability, prosperity, and longevity.

By exploring alternative social and cultural scenarios, we can consider how different values, beliefs, and practices might have led to different forms of government, economic systems, and social hierarchies. We can also reflect on the importance of cultural diversity and the potential for creating a more inclusive and just world.

Examples and Case Studies

The Great Divergence: This term refers to the period between the 15th and 19th centuries when Western Europe began to diverge from the rest of the world in terms of economic and technological development. Some historians argue that this divergence was due to a combination of factors, including geography, institutions, and culture.
The Industrial Revolution: This period of rapid technological innovation transformed the world, leading to unprecedented economic growth and social change. Some historians argue that the Industrial Revolution was made possible by a unique combination of factors,

including the availability of coal, the development of the steam engine, and the rise of capitalism.
The Cold War: This period of geopolitical tension between the United States and the Soviet Union shaped the second half of the 20th century. Some historians argue that the Cold War could have ended very differently, with potentially catastrophic consequences.
Conclusion

By exploring alternative scenarios for the rise and fall of civilizations, we can gain a deeper understanding of the complex interplay between environmental conditions, technological innovations, and social structures. This exercise in imagination can also help us to appreciate the fragility of civilizations and the importance of making wise choices to ensure a sustainable and equitable future.

Subtopic 4: Lessons from Counterfactuals:
Reflecting on the insights gained from exploring alternative histories, considering the resilience and fragility of civilizations, the role of chance and human agency, and the potential for shaping a more sustainable future.

(Research: Philosophy of history, future studies, resilience theory)

Introduction

The exploration of alternative histories, through counterfactuals and "what if" scenarios, offers more than just an entertaining thought experiment. It provides a unique lens through which we can examine the resilience and fragility of civilizations, the intricate dance between chance and human agency, and ultimately, the potential for shaping a more sustainable future. By reflecting on the lessons gleaned from these imagined pasts, we can gain valuable insights into the forces that shape our world and empower ourselves to make informed choices for the future.

Resilience and Fragility: A Delicate Balance

Counterfactual histories often highlight the delicate balance between the resilience and fragility of civilizations. On one hand, they showcase the remarkable ability of human societies to adapt and overcome challenges, even in the face of seemingly insurmountable odds. Consider a scenario where the Black Death had been even more devastating, wiping out 90% of Europe's population. While such a catastrophe would undoubtedly have caused immense suffering and societal upheaval, it's conceivable that human ingenuity and resilience would have eventually led to recovery and the emergence of a new social order.

On the other hand, alternative histories also expose the vulnerability of civilizations to seemingly minor events or choices. A slight change in the course of a battle, a different decision by a key leader, or a minor technological innovation could have dramatically altered the trajectory of history. The fall of the Roman Empire, for example, might have been averted if a series of emperors had made different decisions regarding military strategy, economic policy, or succession.

By recognizing both the resilience and fragility of civilizations, we can develop a more nuanced understanding of the factors that contribute to their success or failure. This knowledge can inform our efforts to build more resilient societies that are better equipped to withstand challenges and adapt to change.

Chance and Human Agency: A Complex Interplay

The exploration of alternative histories also sheds light on the complex interplay between chance and human agency in shaping the course of events. While chance events, such as natural disasters or unexpected discoveries, can have a profound impact on history, human choices and actions also play a crucial role.

Consider a scenario where the Spanish Armada had successfully invaded England in 1588. While this event might have seemed inevitable in hindsight, it was contingent on a series of human decisions and actions, both by the Spanish and the English. Had the English adopted a different naval strategy, or had the Spanish made different logistical choices, the outcome of the conflict could have been very different.

By recognizing the role of both chance and human agency in shaping history, we can avoid falling into the trap of historical determinism, which suggests that the past was inevitable and that the future is predetermined. Instead, we can embrace the idea that we have the power to shape our own destiny, even in the face of unpredictable events.

Shaping a Sustainable Future: Learning from the Past

Perhaps the most important lesson we can learn from exploring alternative histories is the potential for shaping a more sustainable future. By considering how different choices and actions might have led to different outcomes, we can gain valuable insights into the factors that contribute to long-term sustainability.

For example, imagine a scenario where the Industrial Revolution had been driven by renewable energy sources instead of fossil fuels. In this alternative history, we might have avoided the environmental challenges we face today, such as climate change and pollution. By learning from this counterfactual scenario, we can prioritize the development and implementation of sustainable technologies and practices in our own time.

Furthermore, by studying the rise and fall of past civilizations, we can identify patterns and trends that can inform our efforts to build a more sustainable future. For example, many civilizations have collapsed due to environmental degradation, resource depletion, or social inequality. By learning from these historical examples, we can strive to create societies that are more equitable, resilient, and environmentally sustainable.

Conclusion

Exploring alternative histories offers a powerful tool for reflecting on the lessons of the past and envisioning a more sustainable future. By considering the resilience and fragility of civilizations, the interplay between chance and human agency, and the potential for shaping our own destiny, we can gain valuable insights that can inform our choices and actions in the present. Ultimately, the study of alternative histories reminds us that the future is not predetermined, and that we have the power to create a world that is more just, equitable, and sustainable for all.

About Author

Azhar ul Haque Sario is a bestselling author and data scientist. A Cambridge alumnus, he possesses proven technical skills with an MBA, ACCA (Knowledge Level - FTMS College Malaysia), BBA, and several Google certifications, including Google Data Analytics Specialization, Google Digital Marketing & E-commerce Specialization, and Google Project Management Specialization.

With ten years of business experience, Azhar leverages his expertise to deliver insightful books. He has written and published an astounding 2810 titles and was awarded the record for the maximum Kindle editions and paperback books published by an individual author in one year by Asia Books of Records in 2024.

ORCID: https://orcid.org/0009-0004-8629-830X
Azhar.sario@hotmail.co.uk

www.ingramcontent.com/pod-product-compliance
Lightning Source LLC
Chambersburg PA
CBHW031345230926
51848CB00002B/393
* 9 7 8 3 3 8 4 6 9 9 7 8 7 *